Developmental Disabilities

A Simple Guide for Service Providers

Mac Griffith, Ph.D.

COPYRIGHT

Also by Mac Griffith

Lyric River, a novel

Strong Hills: Tales from the Mountains

Dedication

This book is for my teacher, for my boon companion of ski slopes and trout streams and nights dodging falling stars above timberline, and for my daughter—all of whom are Carly.

About the Author

Mac Griffith, Ph.D. has been a clinical psychologist in private practice in Summit County, Colorado. He was educated at Baylor University. He was associated for many years with a community-based agency for individuals with developmental disabilities. He was formerly an associate professor of psychology at Sam Houston State University in Texas. He has published a variety of technical articles in obscure professional journals; it is not known whether anyone actually read any of these articles. He has given many lectures and presentations on various professional topics; again, it is not clear whether anyone listened; while some of the audience did appear to be sleeping, the rumor that they had been rendered comatose by the lecture itself has never been definitively confirmed. Mac has also written a feature column for a local mountain newspaper, humor features for the Denver Post, and has published longer essays in the Mountain Gazette, a regional magazine.

CONTENTS

PREFACE

I am more than moderately surprised to be here, writing a long preface to a short book. The preface comes at the beginning but is actually written last; first you have to write the book in order to figure out what it is you think about whatever it is you are writing about.

I did not have a lot of trouble figuring out what it is I think about working with individuals with developmental disabilities. I just wrote down all the stuff I know; therein, of course, lies the main reason for the shortness of the book.

This is not a traditional academic textbook on developmental disabilities. While it covers some of this traditional information in simplified form, I have written in an informal style that aims for both the heads and the hearts of service providers. Working with individuals with developmental disabilities makes demands both on our knowledge and our humanity—we need providers armed with both. I worked for many years as a staff psychologist for a community-based agency for individuals with developmental disabilities. What follows is one result of this work, although I hope not the most important result. Because it springs from this setting, this book is more clinical than academic and is written for an audience more wide than narrow. I hope that the academic foundation, the basis in scientific knowledge, is, though simplified, still sound. I have tried to focus on a few well-established principles and how to implement these principles rather than sifting through the fine print of knowledge. This is natural given that I am a clinician rather than an academic. There is, in addition, another reason for focusing on broad principles in clinical work. It is that the main ideas make most of the difference.

I wrote in this book about the stuff I know now that I wish I had known when I started out. I had in my head the rather simple-minded idea that others just starting out might want to know some of this same stuff.

I am alert enough to figure out that in the grandest sense we are all just starting out. I do not, however, do windows and neither do I do "grand" particularly well—clearly, if I did grand, this would be a big rather than small book and, furthermore, would be full of the sorts of ideas that would solve all the problems of individuals with developmental disabilities. Were I up to this, I would then push right along and fix up world hunger, world peace, and justice for all. Actually, on the last one, I think I would start with immediate justice for me and justice for everyone else on a gradual basis.

The Nitty-Gritty of Problems

Since I am clearly not up to these grand problems, I have worked on smaller and more practical problems that seemed within my reach. These are the kinds of problems that direct care staff who work with individuals with developmental disabilities face. In some cases it is the kind of information families of individuals with developmental disabilities might like to have. College students might like to know some of this stuff in order to add some clinical perspective to regular academic presentations.

I have made it a goal to write plainly, to use simple language rather than fancy words and academic jargon. I expect I have succeeded only partially because I cannot completely escape owning a Ph.D. in psychology—unless, of course, upon reading this book, the good faculty of Baylor University should be so offended that they send out a hit squad to take back my diploma and, in a ceremony both solemn and terrible, rip to shreds my inkblot cards, and drum me out of the profession. Also, I cannot escape the fact that in a former life I was one of those dreaded college professors. I don't know how to explain this except to point out in my own defense that the pay was good and the work not such as to be over-tiring.

I have also made it a goal to make this book as lively as possible. There are jokes in this book, some of which are shamelessly bad. I try to be funny for a couple of reasons. One is a naked attempt to keep you, the reader, reading. Think of the glory involved if I could actually write a book about developmental disabilities that was too interesting to put down. Okay, okay, maybe this is a bit grand, but you have to grant me that delusions of grandeur are a lot more fun than those of persecution. The other reason I try to be funny is simply to keep myself amused. Writing a book, even a little one like this, can be a grind, and the only way I can keep going is to keep myself amused. My daughter is fond of pointing out that most of my jokes amuse no one but me. What she fails to appreciate, apart from my humor, is that keeping myself amused is most of the battle.

I also note that sometimes in these pages I am ornery and contrary. I am not always entirely respectful of all wisdom that is viewed as conventional. Keep in mind that sometimes when I am contrary I just can't help myself. At other times it is my intention to stimulate some argument, some thought, some disagreement—if I succeed, I will be perfectly cheerful. Also keep in mind that some of my opinions may be completely wrong, and I will continue to be perfectly cheerful even if perfectly wrong for the very straightforward reason that perfection continues to elude me.

Perhaps it might be nice at this point to sketch out where this book is going. I will try to do it carefully so as not to give away too much information about major plot twists, surprise endings, or the part where I get a call from the President of these United States and have to temporarily give up my job as a mild-mannered psychologist and become a secret agent in order to save Western Civilization from the Forces of Evil. I note in passing that I am an equal opportunity secret agent and would be happy on request also to save Eastern Civilization; should this become necessary, the President has my phone number. However, before outlining the book…

A Special Note on Terminology

Mental Retardation (MR) vs. Intellectual Disability (ID)

To this point, I have used the term developmental disabilities, which is widely accepted and without special controversy. Chapters Two and Three address issues of terminology in more detail. The field is currently in a state of flux over the use of the terms **mental retardation (MR)** and **intellectual disability (ID)**. MR is the older term, and an important movement is afoot to replace this term with ID, for a variety of reasons. We are in a period of transition on this terminology. Without taking up the specific arguments, I simply note that such transitions happen periodically and are often accompanied by heated argument; I don't have much interest in the heat, but the arguments can be instructive.

The current argument for changing terms was given significant impetus by the American Association on Mental Retardation, which demonstrated its commitment to the new terminology by recently changing its name to the American Association on Intellectual and Developmental Disabilities (AAIDD). The American Psychiatric Association (APA) still uses the term mental retardation (although this could change in the future). It will take a while for all this to sort itself out. Scholars have to make their arguments and counterarguments. In addition, the matter is complicated by the fact that there are so many forces at work: scholarly outfits like AAIDD and APA, differing practices in other countries, organizations that try to maintain international classification systems, insurance companies, and state and federal laws. It will clearly take some time to get all these cats herded in the same direction, whatever that direction turns out to be. To illustrate, the AAIDD is at pains to point out that ID is synonymous with MR. That is because, in fact, it mostly is and also because many state and federal laws regarding program eligibility still use the MR term. So, for example, the AAIDD quite rightly has no desire to change the diagnosis to ID and then have the state refuse funding because its law only provides assistance to individuals called MR. The fact that the definitions of MR and ID are basically the same serves to illustrate also that mostly what we are changing here is the terminology. Battleships don't turn on a dime; neither does an interlocking system of scholarly organizations, state and federal laws and agencies, international classifications systems, etc. Lest we forget, there is also in this complicated system the actual individual being served. And their actual families. And their actual service providers.

My concern in this book is more with these actual individuals, and I am mostly content to leave the above complications to my betters. In brief, one reason AAIDD and others want to change from MR to ID is the stigma problem. MR has been around long enough for people to commonly use retarded as a casual insult in conversation. The term MR was introduced in part many years ago to address this same stigma problem that was associated with previous medical terms like idiot and moron. I address this issue particularly in Chapter Two, where I argue that human nature in this regard could stand some improvement. This line of argument suggests that the lifespan of ID may also be limited—it, too, may also be turned into a derogatory term with the passing years. You can also argue that if changing the term buys some of those actual people some relief for a while from having a dual diagnosis—in this case, a diagnosis that can simultaneously be both a condition and an insult—then that's not all bad.

Again, there are complicated issues here that I am mostly content to leave to others. For example, AAIDD hopes to get people thinking more about a zillion ways to improve the adaptive functioning of these actual people. This is unequivocally a good thing, especially in this age of technology where a computer can beep at the bus stop for my doctor's office, and other good ideas of this sort. Many improvements await, but, on the down side, it's hard for me to picture an adaptive device, or adaptive training, that will allow the individual with MR/ID to pass college calculus. Improvements yes, miracles not so much. Garrison Keillor talks about his fictional town of Lake Wobegon, "where all the children are above average." It would be an actual miracle if some brain scientist came to us tomorrow with a medicine that would make all the children (and adults) "above average" citizens of Lake Wobegon. If this happened, we would, paradoxically, no longer feel much need to be swept up in arguments about terminology. In the meantime, actual people are struggling for small improvements in their lives.

Maybe some good will come from changing terminology. I don't know. Since I am full of contradictions, I am both skeptical and hopeful. **For this book, I will follow the practice of other writers caught in this transition period and use the MR/ID abbreviation.** I even ran across one journal article where the author used this device in meticulous alternating form—MR/ID was always followed by ID/MR. Too complicated for me, but an admirable intent.

I also note that transitional periods create more opportunities for learning, for sharpening ideas, than are present in quieter times. The Irish like to say, "May you live in interesting times," but no one has quite figured out whether this is blessing or curse.

Chapter Descriptions

Chapter One considers the nature of intelligence and some of the problems we have in figuring out what intelligence is. This chapter does not actually take up the question of MR/ID directly. Rather, it comes at the question in a sneaky fashion by raising the possibility of creating about six billion Einsteins and Shakespeares. In such an imaginary and smart new world, the distribution of intelligence test scores would be changed dramatically and those of us who thought we were pretty smart might be in for an adjustment or two.

Chapter Two addresses in more specific ways attempts to measure intelligence and define MR/ID through a combination of measures of intelligence and measures of adaptive functioning. Some of the stuff in this chapter is a little technical, and I bravely run the risk of dropping the reader into a coma. I persist because these questions are important in deciding who gets labeled with MR/ID and how, and who gets services, and who decides all of the above.

Chapter Three is much safer stuff. It describes the characteristics of individuals with MR/ID and what to expect from these individuals in very concrete terms. I talk about such things as limited vocabulary, problems with generalization, academic deficits, etc. Also included in this chapter is some traditional statistical information on MR/ID as well as a brief discussion of causes of MR/ID.

Chapter Four concerns itself with problems of dual diagnosis. Many individuals with MR/ID also have other mental disorders. The diagnosis of these other mental disorders is often tricky for many reasons. I discuss the sometimes controversial use of psychotropic medicines for individuals with MR/ID.

Chapter Five is about problems of competence and responsibility as regards individuals with MR/ID. Do individuals with MR/ID have exactly the same rights and freedoms as everybody else? Are they competent to get married, to manage money, to go to jail? When should society intervene and restrict freedoms? When should society let the individual stand or fall on his own? What is the proper role of institutions? Of group homes? Are all institutions bad? I don't know the answer to any of these questions, but I have managed to puzzle out one very important principle about all these questions: be very skeptical of anyone who says they do know the answers—beware of Trojans bearing gifts and true believers bearing The Answer.

Chapter Six takes up behavior modification procedures. I discuss in an elementary fashion such topics as positive reinforcement and punishment procedures. I point out that these are very powerful procedures for changing behavior but that most of the time we are able to use them only in a relatively weak form and that this is mostly a good thing in order to avoid dictatorship. Special emphasis is placed on the use of positive reinforcement procedures and the use of activities that are intrinsically reinforcing.

Chapter Seven is about paying attention to people. This is not strong enough—it is about really, really paying attention to people, which most of us don't do but should because it is an amazingly powerful technique that can be used without being a dictator.

Chapter Eight is about power struggles. I am against them and will fight to the death to see that you are, too. Oh, wait. That would be wrong. I am against them and hope you will be, also, after reading this chapter.

Chapter Nine argues the obvious, i.e., there is no place for big language and fancy words and jargon if we hope to communicate effectively with individuals with MR/ID. Individuals with MR/ID mostly do not have big vocabularies. We all know this. So why is it so hard for us to shut up with the big words and speak simply?

Chapter Nine is about how to speak and **Chapter Ten** is about how to listen. I describe some elementary techniques of reflective listening. Being listened to, being helped to tell your story, is also big medicine, and most of the time it is medicine that is left on the shelf. I describe ways of opening up conversations rather than shutting them down. Individuals with MR/ID get ignored a lot, get avoided a lot—they do not get their share of eager and effective listeners to hear their stories.

Chapter Eleven is a tiny little chapter on the slow rate of progress sometimes shown by individuals with MR/ID. The moral to this story is that we may need to slow down our clocks, our expectations, but that change occurs. People improve, but sometimes you have to hang around for years to see it happen.

Chapter Twelve concerns itself with weirdness. Not weird clients or even weird staff. Just weird situations that sometimes pop out of the developmental disabilities system. I talk about a specific individual with MR/ID, stuck for years in exactly the wrong residential setting. And this situation is not exactly anyone's fault, but it is just one of those situations that happens. As long as we have systems, bureaucracies, the occasional weirdness is going to pop up, and everything is not going to make perfect and complete sense.

Chapter Thirteen is devoted to families of individuals with MR/ID. I am not a member of one of these families, so my understanding is limited. But I can ask questions, listen, and learn. And I can ask families to let me form a partnership with them and ask them how best to do this. And if I can work towards being partners rather than adversaries; if I can ask questions rather than giving lectures; if I can look for solutions rather than someone to blame; then perhaps so can others.

Chapter Fourteen is devoted to the stress experienced by staff who work with individuals with developmental disabilities. Burned out staff and staff turnover are large and expensive problems. This chapter proposes some ideas for better stress management.

CHAPTER ONE

EINSTEIN, SHAKESPEARE, AND THE REST OF US

During my very first semester of graduate school, I took a course in intelligence testing. This class met once a week. On the first day of class the professor handed out a reading list. I took a look at this list and thought to myself: this is a lot of reading for one semester, but everyone said graduate school was going to be tough. I spent the week between the first class and the second busily engaged in such scholarly pursuits as softball and hanging out in places where female students might be hanging out. I had in those days an abundance of energy and devoted these energies abundantly to these important tasks. I knew that sooner or later I would need to get around to that formidable reading list, but it seemed to make the most sense not to waste the fine fall weather. There would be plenty of time to read about intelligence later in the semester when the weather turned cold. In any event, I was generally of the school of thought that later is preferable to sooner whenever work rears its ugly head.

This particular class was an evening class. I vividly remember the softball game that occupied most of my attention the afternoon before the second class. I know that I felt moderately virtuous for leaving the game early so I could be almost on time to class. Upon arriving at class, I discovered a couple of disturbing facts. It developed that the formidable reading list was not intended to be read over the course of the semester. This was the reading list for the week. Before I was able to make peace with this unhappy news, it further developed that the professor was actively engaged at that very moment in asking questions of members of the class, picked pretty much at random, about the contents of the reading material. As I listened to a couple of these questions, it became clear to me that none of the questions were going to be on softball or good places to meet female college students.

At the very moment that this unhappy news was sinking in, my name was called. The professor asked me to define intelligence. It came to me that I had just been picked off of first base. I am generally philosophical about situations of this kind. Specifically, my philosophy is that when being smart is out of reach you can always fall back on being a smartass.

Intelligence—a Definition that Kind of Works

I informed the professor that intelligence is what intelligence tests measure. Imagine my surprise when this turned out to be pretty much the right answer. This was one of those times when being a smartass paid off; this turned out to be good for my grade but bad for my character.

Eventually I got around to reading at least some of the books and articles on the reading list. While there are many learned definitions of intelligence, I have yet to come across a one-sentence definition of intelligence that tells me much more than the answer I blurted out in class in a moment of panic. It is possible to describe the typical kinds of things that are measured by many tests of intelligence although this does little more than take us back to the circular answer I gave my professor. Most tests of intelligence include items having to do with verbal skills. This might involve defining vocabulary words. It might include items that seem to reflect abstract verbal reasoning; the individual might be asked to describe how a plane and a boat are the same. Most intelligence tests also include items that measure "seeing and doing" skills. The individual is asked to put blocks together to match a pattern. The individual might be asked to put puzzles together or find his way out of a paper and pencil maze. Tests often also include some short-term memory items like repeating a string of numbers or remembering various visual symbols.

Again, the alert reader will have noticed that all I have done is to describe the test as opposed to providing any real definition of intelligence. Part of the reason we cannot define intelligence very well is tied up with the fact that intelligence reflects brain activity (in rather mysterious ways), and the brain is a bit of a black hole in biology. The heart, for example, is a nice, simple organ, and we understand it so thoroughly that we can swap it around from one body to another, and we are even tinkering with making artificial ones that one day may work as well as the original issue. We are nowhere near to making an artificial brain. Trying to figure out how our brains work leaves us pretty much baffled. We are left in the awkward and embarrassing position of having brains that are too complicated and sophisticated for our brains to figure out.

IQ Tests—Kind of Useful

Fortunately, it turns out that it does not matter terribly that we cannot define intelligence very precisely. Whatever it is the tests are measuring turns out to be pretty useful. The tests allow us to make effective distinctions between people, and they do this within an acceptable margin of error. They give us useful information about who is likely to succeed at calculus and who may stump their toe over the multiplication table.

It can also be helpful to consider what intelligence clearly is not. It is not creativity. If you think defining intelligence is difficult, try defining and measuring creativity. Intelligence is not motivation. There are tons of smart kids who are lazy, who prefer softball to reading lists. Intelligence is not character. Smart people do not necessarily lie less or perform more acts of kindness. IQ scores don't tell us about the individual's capacity to love or be loved.

Intelligence—Frighteningly Relative

You do not have to puzzle long over the meaning of intelligence before you come to an appreciation of how relative all this is. Compare your own personal (and undoubtedly very respectable) IQ score with a couple of plain citizens off the street. Pretend we could fetch in, say, Einstein and Shakespeare. Our own (very respectable) IQ's might look a little puny. Happily, we rarely have to make such discouraging comparisons.

But take a moment and imagine the possibilities. The teacher has asked that you and little Albert E. stay after class and do an extra credit assignment. The task is something as ridiculously simple as computing the weight of the sun. And, just to make this even more an intellectual trifle, you can use your calculator. Little Albert E. jumps up and heads for the pencil sharpener while you scratch what you once thought was a pretty good head and realize that since Albert has already sharpened the pencils there is no other part of this project you can help with.

This is not a pretty picture. It gets worse the next day when the teacher asks if you and William S. would mind staying after class and running up a little sonnet to commemorate the 30th anniversary of the founding of Der Wienerschnitzel. Little Willy S. remarks that he prefers iambic pentameter, and you volunteer to sharpen up a couple of pencils.

A person could feel pretty stupid after an experience like this. A person could feel frustrated. Angry. Insecure. A person could get depressed and lose all confidence in their abilities.

Brave (or Scary) New World

It would certainly be discouraging to have to try to compete with Shakespeare and Einstein. Happily, we rarely have to do this. But what if the normal distribution of intelligence scores were to change dramatically? As things are now, intelligence test scores follow a bell-shaped curve in which the lower one to three per cent of the population is defined as representing MR/ID (Abbreviation Reminder: MR/ID=Mental Retardation/Intellectual Disability). The extreme opposite end of the curve, of course, includes that minority of individuals like Shakespeare and Einstein. The great statistical majority of us muddle along somewhere in the middle.

Think of the difference to the world Shakespeare and Einstein have made. Think of how wonderful it would be if we had more Einsteins and Shakespeares. Perhaps we could find a way to get a few more of these intellects. Say, three billion Einsteins and a like number of Shakespeares. This would have a profound effect upon the world. Think of the advances in science and art. There would also be some significant effects on the distribution of intelligence test scores. Whatever this distribution actually looked like, the practical effect would be to make most of us feel rather slow.

Incidentally, the creation of more smart people is at least theoretically possible through cloning, the exact and asexual reproduction of particular genetic combinations taken from a single cell. The rest of us can only hope the biologists do not figure out how to make this practical.

The Formerly Mentally Advanced

If this smart new world were created, then most of us would soon find we were a touch slow. The smart people would invent lots of new jobs, none of which the rest of us would be able to figure out how to do. Seeing us lag behind, the smart people would likely set about to study us. Once you have commenced serious study, then a label is obligatory. How about the Formerly Mentally Advanced? We would get a file with FMA stamped beside our name. It would probably be discovered that, contrary to prevailing opinion, the FMA's <u>can</u> learn although our rate of learning is slow as molasses. The Shakespeare clones, of course, would probably find a better simile than molasses. We FMA's are likely to perform best at jobs that are relatively routine and structured because we are easily confused. The FMA should not be given tasks that require too much abstract thought because they think concretely. Incidentally, "concrete" as a description for thought could certainly benefit from some improvement by one of the new Shakespeares. Concrete thinking does not refer to thoughts the individual is having about building materials. Alas, neither does it suggest that the individual's thoughts are weighty and thereby full of significance. It just means literal and limited.

The smart people would likely describe us FMA's as being easily frustrated. It is certainly easy to see how we would be frustrated since, in this scary new world, it would now be considered the merest of child's play to develop a mathematical equation that describes the bending of space by matter or, slightly more challenging yet, to actually understand the instructions that come with those children's toys labeled "Some Assembly Required."

Even in the world as it is, without six billion geniuses running amok, I still get frustrated fairly often. I get frustrated because the world is too damn complicated. There are too many things to do, too many things to remember to do, too many things I should have done last week but kind of forgot and didn't really want to do anyway, too many rules, too many forms, and too many rules for filling out forms.

All of this talk of Einsteins and Shakespeares is, of course, by way of illustrating that intelligence is relative. Consider the courage required to express an opinion in a room full of Shakespeares and Einsteins. Imagine that some of these people got together occasionally with you to make plans and help resolve important problems in your life. They might decide to call such a meeting a staffing. Were I the person being staffed, I would think it foolish to actually have an opinion of my own in the presence of the truly smart. I would smile and nod agreeably while groping desperately for some clue as to what they might be talking about. I expect they might infer that I had a poor self-concept and problems with excessive dependency. If I were particularly pathetic, they might decide I was suffering the effects of institutionalization. Whatever they decided, who am I to argue?

The fact that intelligence is relative serves as a reminder that we should look for ways to distinguish ourselves that do not depend upon an IQ score. It was Martin Luther King's wish that his children be judged by the content of their character rather than the color of their skin—this might not be a bad place to start.

So Just Remember This

> We can measure intelligence better than we can define it.

> For many practical purposes, it is useful to define intelligence as what intelligence tests measure.

> These practical purposes include providing broad predictions about future academic and vocational performance.

> Intelligence is **NOT** creativity or character or kindness or other really important stuff.

> Intelligence test scores follow a normal distribution in which individuals with the lowest one to three per cent of scores are typically classified as having MR/ID (more on this later).

➤ Intelligence is relative. Imagine a world in which you were forced to compete with billions of Einsteins and Shakespeares. This is a good reason to be humble and empathetic.

➤ In this imaginary world, you might be labeled and studied and "cared for." You could easily feel pretty stupid, have low self-esteem, and be frustrated a lot. This is one pretty good reason to treat others with dignity.

CHAPTER TWO

MR/ID: CALL IT GLYFF

Everybody sort of knows what MR/ID (Last Abbreviation Reminder: MR/ID=Mental Retardation/ Intellectual Disability) is. As I wander about in this chapter, I will slip in the various technical aspects of the definition. I will certainly hold this technical talk to a minimum to keep from confusing you, the reader, and, more importantly, me, the writer.

Everybody pretty much knows that MR/ID has to do with intelligence, and clearly some people have more intelligence than others although things get a little murky when we start trying to sort out exactly what intelligence is. Think about it this way. If you and I were discussing some issue and we disagreed, and no matter how patiently I explained my position you continued to stubbornly think that you were right, then, as anyone can plainly see, intelligence is that characteristic that I possess and you lack.

This definition of intelligence works for me, but there are those who would argue that my approach is a bit subjective. Early attempts to define intelligence jumped off from exactly this point. A hundred years ago, psychologists began chasing after the problem of objectively defining intelligence. Then, as now, everybody "sort of" knew what intelligence was, but no one had a good way of objectively sorting people into more intelligent and less intelligent groups.

Think Fast

One early attempt at this was to try to get at the notion of something we might call brain speed. This approach was associated with the work of an Englishman named Galton. Experiments trying to measure brain speed were devised. You create a panel with a light and a button to push. The subject of the experiment is asked to rest his hand on a certain spot and as soon as the light comes on his job is to push the button as quickly as he can. You carefully record the amount of time it takes him to push the button. You proceed to measure the response times of thousands of subjects with this and similar tasks. Subjects are taken from all walks of life with the expectation that the fast-witted (the doctors, the lawyers, etc.) will have faster response times than those we have reason to think are not so intelligent.

This was a very clever approach to measuring intelligence. If some sort of brain speed factor were at work, this would have the advantage of avoiding possible bias introduced by educational or cultural advantages. We have chased in various forms over the years this goal of a relatively "pure" measure of intelligence that can be reliably obtained and that is not all mixed up with other stuff like cultural advantage and disadvantage. It has been a merry chase, but this is one fox that always gets away. It turns out we can measure very reliably how long it takes to push a button when the light comes on—the only difficulty is that these measurements are not reliably related to anything we would ordinarily think of as intelligence. Professional rocket scientists manage to make about the same score on this test as professional ditch diggers.

Wait—Not So Fast

At about the same time as these early attempts to measure brain speed, a Frenchman named Binet set out to measure intelligence using a very different approach. Binet recognized that some children were being sent to institutions for those with mental deficiencies, and he also recognized that deciding which children went and which did not was a somewhat haphazard process. Like Galton, he also wanted a more objective approach to sorting people into groups. However, with the wisdom of the French, Binet was not so obsessed with being pure. He devised a test that was more culturally loaded—it included more complex questions about the kind of knowledge that is acquired in school and at home and in other cultural settings. He administered this test to groups of French school children and then compared their test scores to how their teachers rated their overall intelligence. He eventually developed a test that produced a fairly good fit (or correlation) with the judgments of the teachers. Binet found this test to be useful in providing a more objective way of, among other things, deciding who should go to these institutions and who should not.

This approach to testing, as opposed to the brain speed approach, eventually won out. Binet's work was transplanted to the United States, and a revision of his test came to be known as the Stanford-Binet. Subsequently, a man named David Wechsler began developing similar tests, and Wechsler's work laid the foundation for a series of tests that are the most widely used tests of intelligence today.

At Least Be Objective

Both the Galton approach and the Binet/Wechsler approach to measuring intelligence are objective. This means that it does not matter who gives the test; as long as they are adequately trained, every test giver will come up with substantially the same score for the individual being tested. This gets us out from under the problem of subjective judgments; without the test as a guide to judgment, then one person's judgment of another's overall intelligence may or may not show much agreement with the next person's judgment because they may use very different subjective criteria of what constitutes intelligence.

The Galton method gives us an objective number, a reliable number, but the number does not tell us much. The Binet/Wechsler approach gives us a reliable number that also tells us something. Knowing someone's score on a Binet or Wechsler test also gives us a pretty good idea about what that individual's school achievement will be like, about what kind of job they are likely to have, about how much money they are likely to earn, etc. The prediction is by no means perfect—there can be lots of exceptions—but the statistical relationship is fairly strong. These tests give us an objective score that is statistically related to some things with practical importance like school achievement and vocational success.

Achievement Sneaks into the IQ Test

We pay a bit of a price for our reliance on this kind of intelligence testing. When we give up the brain speed approach, we end up with tests that measure a combination of ability and achievement. For example, most tests of intelligence ask people to define vocabulary words and to solve math problems (among other things). Clearly the size of your vocabulary and your ability to solve math problems depends both on your ability and a host of other factors such as educational opportunities, motivation, effort, etc.

Critics can then argue that intelligence tests are biased in favor of those with the best family, educational and cultural opportunities. Arguing up these questions can take us up some complicated mazes. I think the criticisms of intelligence tests are partly true; at the same time, I think the tests are useful and better than any of the alternatives. I especially think that the tests represent an improvement over subjective, unguided judgments—this is the problem that Binet set out to solve, and we are better off with his solution than without it.

And the Numbers Are

Intelligence tests are constructed so that they define the average and most frequent score as 100. About 50 per cent of people score between 90 and 110. About 95 per cent score between 70 and 130. Only about two to three per cent score less than 70 or more than 130. Typically individuals with MR/ID are in this small group that scores below 70 on standard tests of intelligence.

It helps to know exactly how you get your IQ score. You take the test under carefully standardized conditions; this means examiners are trained to give the test in the same way every time and to give the test as much as possible exactly like other examiners. The examiner counts the number of answers you get correct. She then goes to a table of norms and looks up how you compare to a very large sample of individuals in your same age group. This table translates the number of answers you got right into an IQ score.

The Developmental Years

If your score is in the neighborhood of 70 or less, then you may be diagnosed with MR/ID. MR/ID is a term that does not always have exactly the same meaning as developmental disability although, in common practice, these terms are often used interchangeably. Both MR/ID and developmental disabilities are said to have their onsets in the developmental years (before adulthood, which is often defined as age 18). The notion of developmental disabilities includes MR/ID but is a bit broader term that may also include some other conditions in which the individual does not have MR/ID. For example, autism is a condition with childhood onset, but not all individuals with autism have MR/ID; therefore, an individual with autism is always considered to have a developmental disability but may or may not also have MR/ID.

This distinction sometimes also reflects some differences in public policy from state to state. Some states have a developmental disabilities system that includes individuals with MR/ID and also other individuals with autism, cerebral palsy, etc. who do not have IQ's below 70. This kind of system serves more people, and serving more people costs more. It should also be noted that the needs of an individual with cerebral palsy and an IQ of 105 are very different from the needs of an individual with Down syndrome and an IQ of 55. Other states employ more restrictive systems in which the group served consists only of those individuals with developmental disabilities that result in MR/ID.

It should also be noted that some individuals may develop very low intellectual functioning after the developmental years. A person who is 30 and has an IQ of 100 may suffer a head injury in a car accident and subsequently score in the 60's on an IQ test. Usually these individuals are excluded from the developmental disabilities system. In psychiatric terms, these persons are considered to have a dementia or amnestic disorder rather than MR/ID. Often, their scores on the various parts of an IQ test are very uneven, i.e., they continue to show some areas of good mental ability while also showing severe deficits in other areas. There is a higher likelihood that some of these individuals may show some recovery of their intellectual abilities. Again, the needs of these individuals are often very different from the needs of individuals with MR/ID.

Glyff by Any Other Name

We are, at this point, fairly swimming in labels, e.g., MR/ID, developmental disabilities, autism, cerebral palsy, dementia, etc. This gives me the occasion for a diversionary riff on naming things. In the mental health fields, we generally follow the opposite of Shakespeare's rule of the rose. Shakespeare said the rose would smell as sweet by any other name, so it does not matter what we call it. In mental health, we keep trying to change the names of things in the apparent hope that this will remove the not-so-sweet smell of the label. Many years ago, terms like idiot and moron used to be perfectly respectable medical diagnoses. These words gradually became commonly used insults and eventually the names were changed to protect the innocent. I don't think that today's children much use words like idiot and moron. I do know they often call each other retarded, and this is generally not intended as a compliment. As described in the introduction, the American Association of Intellectual and Developmental Disabilities (AAIDD) is currently trying to remove the stigma of mental retardation by changing the term to intellectual disability. It may be that retarded will eventually go the way of idiot and moron and be displaced by intellectual disability. The useful shelf life of the term mental retardation may have expired after a lot of years. As I noted previously, maybe a name-change will spare some good people the pain of suffering mindless insults for a period of time, and this is a good thing. Intellectual disability, as a term, has some things to recommend it. However, consider the result if we used an entirely invented word, say, glyff. As surely as dawn goes down to day, it would not be long before thoughtless people would be using the term glyff as an insult. Teenagers would be calling each other glyff-heads. To paraphrase Mr. Shakespeare, our faults, dear friends, lie not in our technical terms but in ourselves.

How About a Little Support Here?

There is another aspect to the term-changing proposed by the AAIDD (they proposed these changes when they were still the AAMR). The current diagnostic system of the American Psychiatric Association (APA) includes four categories of mental retardation: mild (IQ's of 55-69), moderate (40-54), severe (25-39), and profound (less than 25). (Note that you will see slight variations by different authors on the above number groupings.) The AAMR proposed changing these terms to labels based on how much support the individual needs in living. In effect, we would then refer to individuals with MR/ID who require intermittent supports, limited supports, extensive supports, or pervasive supports. A critical view of the AAMR ideas is that we have taken four old labels and turned them into four new labels. A kinder and gentler view of the new system is that it reminds us to pay close attention to the services (the amount of supports) the individual requires and to focus our efforts on something that can be changed: we usually cannot change an IQ score, but we can try to teach the individual to function more independently, thereby requiring fewer supports.

The Importance of Being Adaptive

This slightly shifts the emphasis from IQ tests as the defining factor of MR/ID to adaptive living skills. Adaptive living skills include such things as whether you can bathe yourself, tie your own shoelaces, write letters, manage money, find your way around town, etc. At first, we mostly measured these skills as a starting point for designing teaching programs for individuals with MR/ID. A lack of adaptive skills was generally thought to be a result of MR/ID. The presence of MR/ID was established through the administration of IQ tests. Some adaptive tests were developed, but these were tests of secondary importance.

Recall that Binet and Wechsler after him had loaded their IQ tests with items that reflected cultural knowledge. They had realized that it was not possible to create a useful test that measured pure ability, so they created tests that represented a combination of ability and achievement. A reaction against this approach eventually developed, especially in the 60's and 70's. Part of this reaction was spurred by studies showing that some minorities in the US scored more poorly, on average, on these tests than whites. The tests stood accused of cultural bias and therefore could be used as instruments of discrimination. Far from being measures of pure ability, the tests were measures of the dominant white culture, and it was small wonder that whites did better on these tests since from birth they were, in effect, taught the test. It should be noted that this argument was employed across the entire range of test scores. A score of 115 might keep a minority student out of a school program for the gifted; a score of 67 might get a minority child unfairly labeled as mentally retarded and assigned to special education classes.

A train of thought that grew out of these real dilemmas was that we should give more emphasis to measures of adaptive functioning. The classic example was that of the minority high school kid who gets that score of 67 on the IQ test, but who demonstrates abundant "street smarts"—the kid who grows up in poverty but who copes with substantial skill within the very different rules of his financially poor environment. The argument was advanced that if we took account of this kid's adaptive skills there is no way we could reasonably call him mentally retarded.

Some version of this debate still rages and likely will continue for the foreseeable future. The issues are difficult and complicated, and I could not resolve them in this short space even if I had the faintest clue how. For our purposes, however, the important thing is that these arguments gave a boost to the notion of adaptive skills. The APA recommended that no diagnosis of mental retardation be made without taking account of adaptive skills. The APA took a conservative position in saying that to get a diagnosis of mental retardation the individual must be below the cut-off point on <u>both</u> IQ test scores and adaptive scores. We run into difficulties, however, when the IQ score is above the cut-off point and the adaptive score is below the cut-off point, or vice-versa.

Labels and Outcomes

Sometimes we don't know how much weight to give to intellectual functioning and how much to give to adaptive functioning. If this were just a technical dilemma, it might be easily resolved. However, what might be merely a technical problem quickly gets wrapped together with the outcomes people are trying to achieve with a diagnosis. Sometimes people want a diagnosis and sometimes not. In the example above, concerning the streetwise teenager, the idea was to take account of adaptive functioning so that this kid is not unfairly stigmatized by culturally biased tests that labeled him mentally retarded and is not held back by sticking him in an undesirable special education program. These days, I seem to encounter more situations that are exactly the opposite. People want a diagnosis because there are benefits and services attached to it. I am more apt to hear the argument that the IQ score may be a little above the MR/ID range, but the individual's adaptive functioning is low and therefore should qualify him for admission into a program. Once we attach dollars and services to a diagnosis, motivations change for understandable reasons.

This problem is then intensified because measures of adaptive functioning are more subjective and less reliable than measures of intellectual functioning. This is so because we often do not obtain measures of adaptive functioning by directly evaluating the individual who might have MR/ID; rather, the common tests are constructed in such a way that we ask questions of someone who knows the individual well, like a family member. We ask a lot of questions about a myriad of specific skills the individual may or may not possess (like street safety). I have seen cases where parents, for example, so want to emphasize their child's strengths that the resulting score terribly overestimates what the individual can do. I have also seen cases where the opposite occurs; in some of these cases, the parent may emphasize the child's deficits in order to qualify the child for programs and benefits.

In either case, the results are skewed. A reasonable argument is that measures of adaptive functioning are more subjective and susceptible of bias, so we should try whenever possible to get a rating from more than one observer. Ideally, this second opinion should come from a developmental disabilities professional who has had the opportunity to become closely acquainted with the individual. The problem is that this is not always possible with individuals who are initially seeking admission to a program.

The emphasis on adaptive functioning introduces more subjectivity, for good or ill, into the system. It can, for example, make more people eligible for services. This is not necessarily bad except I have noticed no rush of politicians scrambling to write more and bigger checks to cover the cost of these services.

All of this brings us full circle to the dilemma with which we began this chapter. The original problem of Binet was to create a more objective way of making decisions about intellectual functioning to avoid the subjective and haphazard assignment of people to one group or the other. The best way of doing this is still through the use of standard tests of intelligence. An emphasis on adaptive functioning reminds us to place information on intellectual functioning within the larger picture of the individual's life, and this is inarguably a good thing. The trick remains one of finding a reasonable balance among the various sources of information so that we can perform the even more difficult trick of arriving at important judgments about the individual that have at least a minimum of both reason and balance. There is no simple answer to how to combine all this information in a way that satisfies a system's reasonable needs for objectivity and the subjective and unique pressures faced by real individuals in the world.

So Just Remember This

➢ **Early attempts to measure intelligence tried to get away from the subjectivity of this concept.**

➢ One approach tried to measure simple things like reaction time. This didn't work out because these simple measures turned out not to be related very well to complex intellectual tasks.

➢ The approach that turned out to be more successful was one that used more "culturally loaded" test questions. The questions involved a mixture of ability and achievement.

➢ This approach was carried forward over many years by people like Alfred Binet, David Wechsler, and many others.

➢ Intelligence tests in the Binet/Wechsler tradition allow us to broadly predict school achievement and vocational functioning. They do so at the price of being more culturally biased.

➢ The tests are objective in the sense that the examiner asks the same questions of everyone in the same way as other examiners.

➢ An IQ test score of 69 or below is the objective "test" definition of MR/ID. MR/ID is a developmental disability because its onset is in childhood. Not all developmental disabilities involve MR/ID; some individuals with developmental disabilities have higher IQ scores.

➢ States vary in terms of how broad their definitions of developmental disabilities are.

> ➤ Terms like mental retardation often get used in hurtful ways because human beings are imperfect creatures. Sometimes it helps to change the term, but it would help even more if human beings became kinder—it takes time.

> ➤ A useful way of re-looking at the traditional categories of MR/ID is to think in terms of the level of supports that will be helpful to these people. This places an emphasis on adaptive functioning within that person's environment.

> ➤ A comprehensive assessment of an individual takes account of both an **IQ** score and adaptive functioning (as well as a host of other things). There is no simple answer to how to combine all this information in a way that satisfies a system's reasonable needs for objectivity and the subjective and unique pressures faced by real individuals in the world.

CHAPTER THREE

CHARACTERISTICS OF MR/ID

Individuals with MR/ID are those who score below 70 on a standard IQ test and who have similar deficits in adaptive functioning. This is an adequate rule-writer's definition but does not paint a very vivid picture of what to expect from individuals with MR/ID.

Vocab Alert

People with MR/ID use a limited number of words and use simple rather than complicated words. So, if you want to have a productive conversation with Ms. Smith, who is an individual with moderate MR/ID, go thou and do likewise. In this paragraph alone, Ms. Smith may not know the meaning of limited, rather, complicated, productive, conversation, individual, moderate, thou, likewise, and paragraph. In ordinary conversations with Ms. Smith, don't use any of these words. Ms. Smith probably does know the meaning of mental retardation although she may wish she did not.

What's the Rush?

Individuals with MR/ID often think more slowly. So speak slowly and allow for longer pauses in the conversation. Be patient about repeating things as necessary. Because someone thinks slowly does not necessarily mean they are going to arrive at the wrong answer. It is not always helpful to rush in and provide an answer for someone who is thinking slowly; there are situations where this can be as rude as finishing a thought for someone who has a speech impairment. Also, it is nice to remember that individuals with MR/ID, like individuals who can't see, are not deaf. Shouting does not increase comprehension but usually is effective in increasing aggravation.

Be Specific

Individuals with MR/ID have trouble with abstract ideas. One of my favorite expressions, whenever I manage to do something right, is "even a blind hog sometimes finds an acorn." This is too abstract for many individuals with MR/ID. This does not mean the idea is too difficult, but it is expressed too abstractly. Instead, say, "I get things wrong a lot, but I got this one right." Keep things specific. The beauty of abstract ideas is that they can be generalized; they can be applied across a wide range of situations that may seem to have little in common; this beauty is lost on many individuals with MR/ID. Essential similarities and differences may be missed. Mr. Jones works on a janitorial crew and has learned not to start vacuuming the halls until six o'clock. On a holiday, when the office building is empty, he stubbornly refuses to start vacuuming before six because he has not understood the reason for the rule.

Individuals with MR/ID learn more slowly. Mr. Jones is given specific and detailed instructions by his supervisor on cleaning out the light fixtures. The supervisor is careful to make the instructions detailed because he knows Mr. Jones has never done this before. The supervisor leaves for another part of the building, and Mr. Jones makes a mess of the job because he could not learn that many new things that fast. Another way of saying this is that Mr. Jones has a poor memory although it should be noted that Mr. Jones could eventually memorize how to clean light fixtures if given sufficient practice trials.

I'm Not Getting It

Individuals with MR/ID have a higher frequency of speech articulation problems. It can feel miserable when you can't grab the faintest clue about what someone is trying to say to you. It becomes powerfully tempting to smile and nod and pretend you understand. It becomes powerfully tempting to find an excuse to end the conversation so you can escape this miserable situation. Rather than escaping, it is better to say, "I'm sorry, I'm not getting it." It is better to ask Ms. Brown to show you what she means. It is better to find someone who has lots of experience with Ms. Brown's speech patterns to translate and to remember that over time you, also, can begin to understand Ms. Brown's speech. In the midst of this miserable situation, it is also helpful to remember that it may be even more miserable and discouraging for Ms. Brown. The odds are that she is trying as hard as she can to speak clearly, would like to be understood, and takes no particular pleasure from the fact she cannot do a simple thing like form words clearly.

Oops

In addition to problems with forming words, individuals with MR/ID often have difficulties with forming other motor movements. There is a higher frequency of cerebral palsy, spinal problems, problems with muscle tone, etc. Physical movements may be clumsy, lurching, and awkward. There may be lots of coffee spills. The nursing department is kept busy with scrapes and bruises from falls. For direct care staff, it is important to remember that bathtubs and stairs and kitchen stoves can become danger zones. I remember my first experience of walking down the hall crowded with individuals with MR/ID. There were always near-collisions that left me unsettled until I figured out my mistake. In crowds, in schools or airports or sporting events, it is remarkable how people automatically adjust their physical spacing between each other so as to minimize collisions. In navigating the crowded hallway at work, I naturally expected that Ms. Brown would move slightly left and Mr. Harrison would move slightly right and I would pass safely between them. Instead, Ms. Brown lurched right, Mr. Harrison spun left (carrying a cup of coffee), and the three of us were on a collision course towards becoming better acquainted.

Problems—It's Just One Thing after Another

Individuals with MR/ID may have short attention spans. There is evidence that there is a higher frequency of attention deficit disorder in individuals with MR/ID. This is one of those findings that I think is true but not as simple and straightforward as it might appear. I took a fair amount of statistics when I was in school. I did okay until the courses began to reach an advanced level of difficulty. Then I got lost, gave up, and quit paying attention. Much of the information to which individuals with MR/ID are "inattentive" is not geared to their interests and abilities.

Individuals with MR/ID experience other mental problems (e.g. depression, anxiety, etc.) at a higher rate than the rest of the population. Some estimates are that this higher rate may be two to four times the amount of "other mental disorders" than the population as a whole. These problems are not defining features of MR/ID, but they occur frequently with it, and it is not terribly difficult to see that MR/ID does not provide a head start on life.

Individuals with MR/ID sometimes act wrong in social situations for any number of reasons like not knowing the rules or being nervous or having poor impulse control or just being plain ornery. So maybe Ms. Black shouts and throws her salad across the restaurant because she wanted the Italian dressing and got cruddy old ranch dressing instead—most people don't do this in restaurants. Mr. Ellis, sitting beside Ms. Black at the table, yells and hits Ms. Black in an attempt to teach her not to make a scene in restaurants. This kind of thing certainly happens and is always regrettable. However, it is important to keep some perspective. When I was in college, in Texas, we used to drive over into Trinity County because it was "wet" and the college was in a "dry" county. I don't remember the name of the bar we went to, but I recall it had sawdust on the floor, and, for reasons still baffling to me, cardboard egg cartons tacked in a solid layer across the entire ceiling. The behavior of most of the people in this bar (my own self excepted, of course) was such as to make Ms. Black and Mr. Ellis seem angelic in comparison.

Some Numbers, Some Expectations

And now, to round out this picture, another regrettable necessity: statistics. Between one and three per cent of the population are individuals with MR/ID. Of these individuals, 85 to 90 per cent are classified in the mild range of MR/ID. The remaining ten to 15 per cent make up the moderate, severe, and profound categories, and there are fewer individuals in each respective category as the IQ score decreases. There are more males than females in each category.

Almost all individuals with MR/ID should graduate high school, most with special education assistance. It is not unusual for individuals with mild MR/ID to develop academic skills around the fourth to sixth grade level. I provide these estimates cautiously as there can be a lot of variability in school achievement, and I certainly do not mean to suggest that a particular level of achievement is all that can be "expected." With the achievement levels suggested above, the individual has some reading skills (and may often read for enjoyment), some writing skills, and generally addition and subtraction skills. In my personal experience, I rarely encounter individuals with mild MR/ID who have good multiplication and division skills. Again, some individuals with mild MR/ID exceed these levels and others do not attain them. Individuals with mild MR/ID are those who are most likely to be able to hold down a community job and live independently with perhaps only minimal supports.

Individuals with moderate MR/ID are unlikely to read much above the second grade level. This means the individual may have some sight words, especially those involving safety and adaptive functioning like street signs. The individual may be able to do some simple counting such that he knows how many quarters are required for the pop machine, but she is less likely to be able to make change. Individuals with moderate MR/ID are likely to require more intensive support in vocational and residential functioning.

Individuals with severe and profound MR/ID rarely develop any traditional academic skills. Usually they are unable to write their names, to recognize letters of the alphabet consistently, or to do any counting. They typically require intensive levels of support from caregivers.

Causes

A good general rule is that we don't know the causes of most cases of mild MR/ID; for most cases of severe and profound retardation, we know the cause to be genetic or related to brain damage. MR/ID can be caused by gene defects, maternal infections during pregnancy, toxic exposures, extreme malnutrition, etc. There are bunches of biological syndromes associated with MR/ID.

In the category of mild MR/ID, there are some biological syndromes, but there is less likely to be a clear cause for the retardation. It is of interest that mild MR/ID is more common in lower income groups while severe and profound MR/ID is spread evenly across income groups. The association between mild MR/ID and poverty is a complicated one since it could involve a mix of influences like access to medical care, nutrition, environmental stimulation, educational opportunities, etc.

Most of the medical syndromes associated with MR/ID, once identified, do not lead to a medical cure. This does not mean the study of these syndromes is unimportant. It just means that most of the interventions that help are those that involve providing environmental supports and education. While there is generally not a medical cure for MR/ID (except in the sense of prevention), proper medical care based on a knowledge of particular syndromes is vitally important to sustain individuals who are often physically fragile and who have a life expectancy, on average, shorter than for the population as a whole.

So Just Remember This

➢ When talking with people with MR/ID, leave your big vocabulary at home.

➢ Give the individual more time to think—slow down.

➢ Keep communication simple and specific rather than abstract.

➢ Take the time and effort to understand even if someone's speech is not clear.

➢ Learn to adapt to unpredictable motor movements.

➢ Expect that individuals with developmental disabilities may have more problems with depression, anxiety, and other mental disorder.

➢ Somewhere around two per cent of the population has MR/ID.

➢ About 85 per cent of this group falls in the mild range of MR/ID. The moderate, severe, and profound groups are progressively smaller.

➢ Adults with mild MR/ID are likely to have academic skills around the fourth grade level. Individuals with more severe impairments have correspondingly lower academic skills.

➢ We don't know the cause of most mild **MR/ID.** More severe forms are likely to be caused by genetic defects or clear physical impairments to the brain.

➢ Many medical syndromes are associated with **MR/ID,** but knowing this does not necessarily lead to a medical cure.

CHAPTER FOUR

DUAL DIAGNOSIS

Problems, like grapes, often come in bunches. People with MR/ID or other developmental disabilities are also likely to have other mental disorder difficulties—hence the term dual diagnosis. In such cases, diagnosis number two is some other psychiatric disorder like depression or schizophrenia.

It used to be that pretty much all the problems that individuals with MR/ID had were attributed to the MR/ID. In a previous edition of the psychiatric manual, there was a diagnosis of mental retardation and another diagnosis of "mental retardation with associated behavioral problems," and these two were pretty much thought to cover it. If you went about screaming or crying or hitting people or hearing voices or just generally being noncompliant, then you got number two. This system had the virtue of simplicity.

It had the vice of blinding people to other psychiatric difficulties individuals with MR/ID might have. It may have caused us to forget the lesson of Shakespeare's Venetian merchant: "If you prick us, do we not bleed? if you tickle us, do we not laugh?" The merchant wanted us to appreciate that he was both Jewish and fully human. It is also possible to be Jewish (or black or brown or white) and MR/ID and fully human.

Recognizing this problem, the APA diagnostic manual was revised and diagnosis number two disappeared. Practitioners were then directed, in effect, to examine all the other diagnostic possibilities that might accompany mental retardation—dual diagnosis. If the individual were crying and withdrawn and had poor appetite and energy, then perhaps he was both mentally retarded and depressed. If all these symptoms were preceded by the death of his mother, then perhaps he had been pricked, had bled, and was fully human.

Psychotropic Drugs as Straitjackets

There was another factor that gave a helpful kick in the pants to this business of dual diagnosis: psychiatrists developed psychotropic medicines that actually worked in a way that was a little more target-specific. In the not so old days, if you were mentally retarded and in an institution and a bit of a problem then you got a bit of Thorazine or Mellaril. These are perfectly respectable medicines that are calming and sedating. And, if you are sufficiently sedate, you are not likely to be much of a problem to staff. An additional benefit of these medicines is that they take away the necessity of unsightly straitjackets, thus providing more opportunity for fashion diversity. The use of these medicines for chemical restraint gave psychotropic medicines in general a bad reputation, and, not coincidentally, eventually unleashed an absolute blizzard of regulatory procedures and paperwork designed to guard against the practice of chemical control of behavior problems.

Individuals with MR/ID have a higher frequency of "other mental disorders" than the general population. There are phobias, obsessive-compulsive disorders, schizophrenias, etc. Individuals with MR/ID suffer from lots of depression and anxiety; it should also be remembered, of course, that individuals in the general population also suffer from lots of depression and anxiety.

Blissful Ignorance or Increased Vulnerability

There are two competing and contradictory arguments that are sometimes heard about MR/ID and other mental disorder. One is the ignorance is bliss argument. The notion here is that individuals with MR/ID are unaware they have problems, unaware there are problems in the world, and simply go about being blissfully happy in their ignorance. This argument bears some similarity to earlier arguments applied to slaves in the South, singing as they went happily to work picking cotton. With a few exceptions, it is baloney.

The more accurate argument is that individuals with MR/ID are more vulnerable to various mental disorders. Most individuals with MR/ID are not blissfully ignorant of their handicaps; on the contrary, they are keenly aware of it and often keenly angry about it.

People with high IQ's talk about stress, caused by the complexity of today's society, and the difficulties of keeping up with the pace of change. A reasonable argument could be made that a simple, farming society might be more congenial to the needs of many individuals with MR/ID. If this is the Information Age, then it's lucky for you if you are very good at storing and processing it and not so lucky otherwise. I note in passing that "storing" and "processing" were once called "remembering" and "thinking" for those of us with older data retrieval systems. From this perspective, it should not surprise us that individuals with MR/ID often have problems with anxiety. I also note that the Information Age may give us some adaptive technology that could make up for some of the headaches of complexity.

Take Depression (Far Away, if Possible)

There are a zillion theories of depression, but I especially like one that seems to zero in on the kinds of problems faced by individuals with MR/ID. This theory takes note of the idea that we often become depressed because of insufficient reinforcement. One instance in which this can occur is when we suffer a loss of a major source of reinforcement—a job, a family member, etc.; individuals with MR/ID lose jobs and perhaps more than their share of family members. An extension of this idea points to the fact that some individuals get stuck in situations where there is not much reinforcement available. This occurs, for example, when you are working for a small company and are aiming for a promotion and then look around and realize that your major competition for becoming vice-president of the company is the owner's son. There simply may not be much reinforcement available to you in this environment. Many individuals with MR/ID are plugging away in situations where there is not that much reinforcement to be had. The other part of this idea about reinforcement and depression has to do with the extent to which the individual is able to attract reinforcers. It is no news flash to point out that individuals with developmental disabilities are not always pretty, not always rich, not always clever, not always highly skilled, and on and on we could go with the "not always" parade.

So it is not a great big wonder that we see a fair amount of depression in this group of folks. It is a great big wonder how much "non-depression" there is. I have not seen much blissful ignorance, but I have seen many individuals with developmental disabilities who have the courage to go ahead and be happy anyway in quite abundant measure. Fortunately, most of us are not required to demonstrate this much courage, although I sometimes think it would be nice were there a requirement to stand and salute in its presence.

The emphasis on dual diagnosis requires us to keep a careful eye out for other problems the individual might be experiencing. If I develop symptoms of schizophrenia, then family and friends and colleagues would help me find professional care. The same applies if I become depressed or anxious. If everyone and their brother and mother are taking Prozac, then perhaps we should consider whether a medicine of this sort might help an individual with MR/ID. I expect that in any event the Constitution of these United States may soon be amended to extend the right to Prozac to all God's children.

Finally, Some Helpful Drugs

This circles us back to an idea we have discussed previously. There is no hue and cry to close medical hospitals for the simple reason that they provide a service that works. The drug companies and the psychiatrists are starting to give us some medicines that work. Institutions for the mentally retarded have a history of not working so it is hard for us to get over the idea that they could ever be good. Medicines for the mentally retarded formerly did not work, so we developed the idea they were bad and wrote all manner of rules to prevent their badness.

In past years, doctors would write prescriptions for extremely sedating medicines to be given PRN (as needed) to individuals in institutions when they became agitated or hard to control. This eventually came to be seen quite properly as being chemical restraint, which was a convenience to staff and which was abusive because it was easier than addressing the substantive problems of the individual. In my state, the regulatory response to this was to forbid PRN prescriptions of psychotropic medicines. So far, so good. But now consider the fact that if you have problems with occasional sleeplessness, you can visit your family doctor and possibly get a prescription for a mild tranquilizer (like Xanax) to use on occasion to help you sleep. Your doctor is likely to specifically direct that you take this medicine only as needed for a variety of medical reasons. Most individuals with MR/ID now have family doctors. They don't visit the institution physician. The family doctor decides to follow the standards of good medical practice she uses with all her other patients and prescribes some Xanax PRN. The current state of the regulations insists that you either get no Xanax or you take it every night, which may not be consistent with good medical practice. Now this regulation itself is not the great biggest of deals; sooner or later some rule writer will figure out a way to fix it; the point is that when it comes to psychotropic medicines we have a hard time figuring out whether it is our job to promote this good thing or fight this evil thing.

Equal Opportunity Help

It is probably a good rule that the same medicines that are available to provide potential benefit to everyone else should be available to individuals with MR/ID. These include medicines for mood disorders, medicines for anxiety problems, medicines for psychoses, medicines for attention-deficit disorder, etc.

These medicines, when used, should be to help the client and not the staff member.

These medicines, when used, should not be used as an excuse for not providing other forms of treatment, such as counseling and other supports. There are too many nots in the previous sentence, but when you are writing Commandments it is hard to avoid using a lot of nots.

Tricky Diagnostic Problems

The use of these medicines needs to be closely monitored. The reason the use of these medicines needs to be closely monitored is because diagnosis and treatment of mental disorders in individuals with developmental disabilities is a tricky enterprise. It is not all that easy in individuals in the general population.

Diagnosis can be tricky because most psychiatric diagnoses depend in major part upon oral communications. We have to do an interview. This is difficult if the individual is severely impaired and has no spoken language. It is difficult if the individual has spoken language, but it is so garbled by articulation difficulties that no one or only a few people can understand it. I note in passing that there is almost always someone who understands these communications (a mother, a brother, a staff member who has worked with the individual for many years), but quite often this other individual is not invited along to the interview. Even when the professional understands the words, he may not understand the message. Many individuals, even at the higher end of the MR/ID spectrum, have trouble with accurately conveying number and time information.

Q: "How long have you been having trouble sleeping?"

A: "A long time." This answer may in fact mean all of last night. Or it might mean two years or two days or anything in between, and it can be damnably hard to sort it out.

The potential for misunderstanding, when inquiring about psychiatric symptoms, is practically limitless. How, for example, do you ask an individual with MR/ID if he has been hearing voices? Most of us would know what this question means because we know we are talking with a psychiatrist and we have heard of hallucinations and so we figure out the meaning of the question from this context. An individual with MR/ID might not know what a psychiatrist is and might not know what an hallucination is and might just think that this is a pretty dumb question. "Of course, I'm hearing a voice. The voice is yours and it is asking if I'm hearing a voice." These are not insurmountable barriers, but they can pose problems that can obscure accurate diagnosis. I try to get around this particular problem by asking the client about his eyes and his ears, how they are working, if there is anything scary or different about how they are working, etc. Also, of course, I try to closely interview staff or others who spend a lot of time with the individual.

If the individual is substantially nonverbal, the problems become that much more difficult. There is a lot we can learn from objective observations. Is the person crying a lot? Is he spending more time alone in his room? Has his hygiene gone downhill? These are all good sources of information, but there are still some problems of differential diagnosis that depend to a great extent upon language.

For example, sometimes people fly into rages because they have a manic disorder and sometimes they fly into rages because they are suffering from post-traumatic stress syndrome. It can be difficult to distinguish between these problems if we can't ask the individual if he has ever been abused, if we can't ask him if he is having intrusive recollections of past abuse, etc.

One moral to all this is that the ideas associated with dual diagnosis are good ones, but the business of making the diagnoses is tricky, and we probably do not make these diagnoses as reliably as we do in the general population.

Helpful Yes—Miraculous, Not So Much

Another moral is that psychotropic medicines are probably getting good enough that we need to cautiously promote their use rather than defend against it. Certainly psychotropics have been associated with some past abuse and can still be subject to misuse and abuse. It is also important to appreciate that psychotropic medicines, even the fabled Prozac, are not sure-fire cures for anything and everything. Would that it were so! I have seen people tried on Prozac who failed to get better and some that got worse. Most of the time, the most we can expect from medicines is some percentage of improvement, whether the medicine is an anti-depressant, an anti-psychotic, etc. An interdisciplinary team that cries out for a medicine and does nothing else to fix the problem is a ship of fools.

Much of the research on medicines for individuals with developmental disabilities consists of case studies so there are no comparison groups to ensure that the favorable effect was actually due to that specific medicine as opposed simply to being given any medicine that, together with extra attention, creates the powerful expectation in the individual that he will get better. This is also a more general problem in research on psychotropic medicines with all groups of patients. In good studies, there are control groups who get a placebo drug, the equivalent of an empty capsule, to ensure that the simple act of being given a medicine is not solely responsible for the patient getting better. With psychotropic medicines, it is sometimes difficult to create a true placebo group because the real drugs often have pronounced effects or side-effects, which make the patients and their observers aware that they are in the "real drug" group. This can then bias the outcome of the study and make it appear that the drug is more powerful and beneficial than it really is.

The issue of side-effects creates another reason for exercising due caution with medicines. Most drugs have side-effects. Psychotropic medicines can create a variety of side-effects such as dry mouth or dizziness or sedation. If the side-effects outweigh the benefits, doctor and patient call it off and try something else. We sometimes have trouble diagnosing other mental disorder in individuals with MR/ID because they cannot tell us a lot about their symptoms. We should also remember that they often cannot tell us a lot about their side-effects. It is important to remember that some psychotropics can make some patients feel miserable in a variety of ways, and if they were miserable to begin with it can be hard to separate the original misery from the misery we have just added on top of it by using the wrong medicine.

This whole business is an elaborate balancing act between costs and benefits. If it sounds complicated, it is. I keep hoping that in the next generation of psychotropic medicines they will come out with a cure for complicated.

So Just Remember This

> Individuals with MR/ID also have a high frequency of other mental disorders, hence the term dual diagnosis.

> Current good practice is to carefully consider the possibility of other problems, like depression, rather than simply attributing all of the person's difficulties to MR/ID.

> This approach was given impetus by the gradual development of useful psychotropic medicines that were targeted at specific problems.

> In the past, there had been too much reliance on broadly sedating methods, which sometimes served as chemical straitjackets. Individuals with MR/ID are not generally "blissfully ignorant and happy;" in

fact, they are more vulnerable to other mental disorders.

➤ They may be more prone, for example, to depression because it is more difficult for them to attract positive reinforcement.

➤ The field has begun to change in the direction of recognizing that individuals with developmental disabilities should be afforded the same kinds of useful treatments as everyone else.

➤ Special precautions should be taken because the diagnosis and treatment of other mental disorders can be difficult for this group of people.

➤ Diagnosis can be tricky because it often depends heavily on spoken communications.

➤ It is a mistake to think of psychotropic medicines as a magic cure for anything.

➤ Unwanted side-effects of medicines are often difficult to assess.

➤ The process of weighing off costs and benefits is a complicated one.

CHAPTER FIVE

COMPETENCE, FREEDOM, AND
AUNT CLARA

When I was a kid, my Aunt Clara often referred to herself as addled. Addled is a nice word. When my aunt was confused, when she was upset, when she could not figure out what in the world to do about something, she referred to herself as addled. Most of the time, I think we are addled on the subject of how much freedom and responsibility individuals with MR/ID should have.

Look closely at the last sentence, and it speaks to the nature of the problem. As long as I break no laws, no one presumes to write about how much freedom and responsibility I should have. If you are writing about me, then you had better write that I will have just as much freedom as everyone else. O.K., I wouldn't mind having a little more freedom than everyone else (and a shade less responsibility), but if I can't have this arrangement, then I will settle for being even up with everyone else.

But individuals with MR/ID are not even up, and trying to figure out what to do about it makes me (and lots of others) addled. A glance at history suggests that generally there has been movement towards the more humane treatment of individuals with developmental disabilities. But even this sentence sticks us back on the horns of the same dilemma. If I, in my generosity, am extending to you more humane treatment, then it follows that I have more power than you do and it also follows that there is something wrong with you that you need treatment.

A History of Rights Restricted

Historically, individuals with developmental disabilities have generally had their rights as individuals restricted in many of the same ways as individuals with mental illness. One difference between these two groups, however, is that it was probably easier to hide many kinds of mental illness from the people who would take from you your rights. In addition, of course, there have always been a wide variety of "milder" mental disorders that did not get you labeled as incapacitated (mild anxieties, depressions, etc.).

Many people are aware from school books of some of the barbaric history of treatment of both these groups of people. They got no treatment or barbarous treatment or were locked away in large and isolated and under-funded institutions where they got a combination of neglect and barbarous treatment. Even when staff were dedicated and kind, they did not have much of an idea what to do that might help, so that eventually it came to be thought that even to be in an institution was itself a form of cruelty.

I note in passing that this was only true for institutions that did not work. In past centuries, hospitals for the physically ill were sometimes horrible places. Eventually, however, scholars made progress in coming up with ideas about physical illness that actually worked (e.g. the germ theory of disease). Doctors began curing more people than they made miserable. As a consequence, rarely does anyone complain today about being "institutionalized" in a hospital on the occasion of having a burst appendix; rather, they demand it as a right and revere it as a privilege. There is no movement afoot to get rid of hospitals.

Humane Treatment

One thread of history has been to move towards more humane treatment of individuals with developmental disabilities. Another thread is to move in the direction of arguing that there is nothing wrong with this group of people that requires treatment, that all they need are "services," that they should have full civil liberties, and that then we should get out of their way and let them stand or fall as everyone else stands or falls. It is somewhere in the cross-currents of these ideas, especially when I begin to try to make them fit the lives of real individuals with developmental disabilities, that I begin to get addled.

Ideas concerning humane treatment of individuals with developmental disabilities start out being fairly straightforward as long as they spring from a logic based upon treatment. It is as if these individuals are "sick," so we as a society owe them certain obligations. They should have enough to eat. And having enough to eat should not be contingent upon acting in a particular manner—this means that crucial edibles like dinner cannot be used as reinforcers to force good behavior out of people. They should have medical care when they are sick. Similarly, they should have good and safe shelter, clothing, transportation, etc. They should be safe in other ways also—they should be free of physical punishment and sexual abuse and emotional abuse.

Clearly the relationship described above is that of powerful doctor to helpless patient or powerful parent to helpless child. The powerful caregivers need to insure that those in their care are not suffering from a lack of essentials like food and shelter. Another part of humane treatment is that we should use our power to enrich the lives of individuals with developmental disabilities. It is not enough to be free of starvation. We should also help individuals with developmental disabilities to have interesting lives that involve such things as recreation, friendships, and family contacts. The life should be safe, and there should be some quality mixed in with it. As long as we are the powerful caregivers, then it is mostly up to us to add the quality to the recipe. The Special Olympics is a wonderful thing, but it has depended to a great extent not on the Kennedy with MR/ID but rather on her smart and powerful sister, Eunice Kennedy Shriver, who helped to organize these events because they are a form of enriching and humane treatment.

More Freedom

Most of us have little trouble thinking that the advances described above are good ones. The good part is that these ideas are aimed in the direction of making the lives of individuals with developmental disabilities as much like the lives of everyone else as possible. Given this foundation, we are then off and running. "Everyone else" does not live in an institution, so let's get people out of institutions. Suddenly we have group homes everywhere— not, of course, without a few fights with the neighbors when they find out who bought the house next door. Given this advance, then we look around and discover that "everyone else" does not live in group homes, either (with the possible exception of college students). If you are an adult it seems more normal that you should live in a house or apartment by yourself or perhaps with just one housemate. So we try to "graduate" people from institutions to group homes to living by themselves in apartments. There is now even a wonderful program to help individuals with MR/ID (and other handicapping conditions) become first-time home buyers.

I am acquainted with quite a number of individuals who have progressed through these graduation steps. Some are happy and some are miserable. Mr. Garrett owns his own modest home and makes better than minimum wage as a janitor at Wal-Mart; he plays on a city softball team and has a number of friends and is delighted with his life.

The Down Side

Mr. Fortson graduated from a group home to his own apartment and a life of excruciating loneliness. He doesn't work because he keeps getting fired and he doesn't care much because his disability checks are enough to get him by. He loves coffee and drinks it all night long until he is so wired up he is ready to jump out of his skin. All the coffee takes away his appetite, and he is rail thin. When he sleeps, it is briefly and usually in the daytime. He sits home and watches TV, and when he cannot stand this anymore he goes to a coffee shop and drinks more coffee and tries to pick up waitresses because he wants a "normal" and "pretty" girlfriend, and he has discovered that waitresses have to talk to you, at least a little bit. Of course, you have to keep ordering more coffee to keep the waitress coming back. This works until he gets so wired on the coffee that his judgment slips a cog and he stammers out some proposal, indecent or otherwise, that then gets him a visit from the manager who asks him to leave. Mr. Fortson then goes back to his apartment and watches TV where the people are all smart and pretty and find true love, on average, about once every nine minutes. While he is watching real life on TV, Mr. Fortson makes another pot of coffee.

To complicate the issue further, I note that Mr. Fortson and Mr. Garrett are individuals with mild MR/ID. The problems become stickier when we are dealing with individuals with more severe delays, individuals with little language, individuals who cannot prepare food for themselves, individuals who are sometimes violent or who have poor control over sexual impulses. This is when we really start to get addled.

This takes us back to a question with which we started. What is wrong with individuals with developmental disabilities? Does whatever is wrong require "treatment?' Does it require "services?"

Competence

A trend over recent years has been to expand the range of activities for which individuals with MR/ID are considered competent. Most people would agree today that individuals with profound MR/ID are <u>not</u> competent to manage money or get married or rear children or to enter into various other legal contracts. This used to be the position that was generally adopted as regards most individuals with MR/ID, even individuals with mild MR/ID. There has been a steady movement to extend more competencies, more rights, and more responsibilities to individuals with mild MR/ID.

Many of these individuals get married. Many have children. Most are considered from a legal standpoint to be independent adults. Some manage their own money. Laws and practices in these areas vary from state to state, from issue to issue.

In Colorado, for example, a woman can secure a tubal ligation in order to prevent pregnancy if she has MR/ID as long as she is also legally judged to be capable of providing informed consent. What constitutes informed consent can be a debatable question. But, if this woman is not judged to be capable of providing informed consent, then it is legally difficult for her to obtain a tubal ligation. There is, here, a bit of a Catch-22 since some might argue that the person who cannot give informed consent might in some cases be the person who most needs the tubal. But it should not be completely surprising that there are some contradictions present in an area where the rules are in a state of flux.

Today, many people consider it thinkable that individuals with mild MR/ID should make their own decisions and have full legal rights. Yesterday, many people would have considered this unthinkable. Tomorrow, will the line be drawn to include individuals with moderate MR/ID, severe MR/ID, etc.?

This works us back to a consideration of what it is exactly that is "wrong" with individuals with MR/ID. This is one of those questions that seems easy but can easily get muddled. When my Aunt Clara was not addled, she was, on occasion, muddled. A comparison of questions involved in MR/ID and in other areas of psychology can be helpful in clarifying the muddle.

Under the current APA diagnostic system, you are considered to have a mental disorder if you are an individual with mental retardation. Definitions of mental disorder, however, are not etched in stone. This is illustrated by the periodic revisions of the American Psychiatric Association's diagnostic manual. In just one example of the changes that can occur, individuals who are homosexual used to be classified as having a mental disorder.

Homosexuals eventually succeeded in "liberating" themselves from this label.

I think we have to beware of "true believers" bearing easy answers. There are limits to the usefulness of ideas. Years ago, physicists beat each other up because one camp found it useful to consider light as a bunch of tiny particles and another camp found it more useful to consider light as waves. A physicist with a sense of humor eventually suggested that light was probably best described as consisting of wavicles.

Does One Size Fit All?

There is a powerful movement afoot to close down institutions. This movement has led to some very nice results. Its emphasis is on achieving a more "normal" life for individuals with developmental disabilities. But maybe there are limits to the usefulness of this idea. All individuals with MR/ID cannot get married and have 2.3 children and live in a house in the suburbs and commute to work in a Ford Taurus.

As noted earlier, we don't want to close hospitals for the physically ill because they work. I think one of the paradoxical things that has occurred in recent years is that institutions have gotten better. The amount of treatment, the humanity of treatment, the effectiveness of treatment has improved. We have better enforcement against abuse; we have better educational and behavior modification programs; we have more focus on enriching people's lives; we have more careful and judicious use of medicines; and, we have better trained staff. Politicians have sometimes been eager to fall in with arguments for shutting down institutions because they are very humane individuals and also because institutions are expensive. Politicians get to be good guys and shut down the institutions, but then they sometimes under-fund community care and take the money they saved by being good guys and build some nice new highways.

I sometimes have the sinking feeling that in twenty or thirty years we are going to rediscover the usefulness of institutions for some purposes. Some radical thinker is going to point out that we have individuals with MR/ID spread out all over the community, and they are lonely and isolated and vulnerable to abuse and not able to obtain the kinds of services they need. This radical thinker may argue that we have scattered them out so as to make them isolated and invisible and that this is a scandal. Examples will be put forward of individuals living in filth in substandard housing. Individuals can be found who have been given control over their money and who, in turn, are being ripped off by unscrupulous merchants or even family members. Individuals can be found who need services but cannot simply go out on the free market and find a doctor or dentist or psychologist or speech therapist who knows much of anything about the problems unique to individuals with MR/ID.

This radical thinker, thirty years from now, may then propose that the clear answer is that we need to centralize these services. We need a group of professionals in one location who know how to provide these services. We need to fund these services. The people who need these services should live in proximity to them, especially since it is such a long commute in from the suburbs in the Ford Taurus. It would be part hospital, part university campus—we could call it, let's see, an institution. There's a fine word. After all, most of the things we truly value in our society—marriage, the family, representative democracy—are referred to as (guess what?) institutions.

None of this is to argue we should go backwards. It is just to point out that sometimes it is hard to know backwards from forwards. Ideas that work in some areas do not work in others. One idea I do have some confidence in is to beware of people who think that their piece of the truth is the same thing as the whole truth.

Stepping Up—Stepping In

I think sometimes we have to look for ways to extend more control to individuals with MR/ID. And I agree that sometimes this freedom is not even ours to extend—I cannot give you what is already yours. At other times, we have to ensure that some individuals do not have some freedoms. We have to be willing to step up and forbid some things. Sometimes doing the one is as hard as doing the other. We are sometimes so successful in preaching about clients' rights that staff are terribly reluctant to set a limit, to take away a freedom.

Mr. Ramos does not have the right to go up to people and pinch them until he draws blood. He has no oral language; he is an individual with severe MR/ID; we are not able to know what he intends by these pinches—perhaps he is only saying hello; but, he still cannot be allowed to do this because we have to act to protect his victims. Ms. Jones cannot be allowed to spit on people at random although I do confess that it might be fun if she could be induced to spit only on people whose names were on a list I would happily supply her. Mr. Harris cannot be allowed to masturbate on Main Street. Mr. James, who is mildly retarded, cannot be allowed to sexually fondle, maul, or rape women with severe MR/ID who cannot testify against him because they have no language to testify about anything, and Mr. James is as cunningly aware of this as any sex offender with normal intelligence. Mr. Massaro cannot be allowed to hit his roommate with a baseball bat because his roommate changed the TV channel.

The Dignity of Risk

There is some behavior we have to stop. There are times we have to decide what is best for people because they cannot decide for themselves or are making bad decisions. Sometimes, for some people, they can decide; they can assume the dignity of risk, and stand or fall on the wisdom of their own decisions. Sometimes those decisions are wonderful and wise. Sometimes they are terrible. I know a woman with mild MR/ID who decided to marry a man with mild MR/ID who physically abuses her. Sometimes she hides her injuries. Sometimes she complains and then relents and becomes an uncooperative witness. All of this is just like a zillion other cases of domestic violence. Lately her husband has taken to abusing her in crafty ways that do not leave marks. I knew she should never have married him, but I have also have known this about others of my friends who have IQ's of 130. She is a free citizen—she can make her choice and reap the reward or the whirlwind. Perhaps, like many of us, she will learn from her mistake. Or, like many of us, perhaps not.

But what about Mr. Ramos who pinches people, perhaps because he has no other language for saying hello. We cannot let others be injured. We cannot let Mr. Ramos be injured when one of his victims smacks him over the head with a chair. This happened once, and Mr. Ramos did not learn from it. Mr. Ramos may be committing an assault with his pinching, but there is no legal consequence that is especially practical. We have to decide for Mr. Ramos what is best for him and set limits. Perhaps we can teach him a better form of greeting; certainly, we will discuss in another chapter ways of doing this. But we should keep clearly in mind that in teaching him a better way and implementing a behavior modification plan to do this we are deciding what is best for him.

Aunt Clara would be addled. I am addled. We are all constantly addled about who decides what when. Sometimes I think we are too paternalistic, and sometimes not paternalistic enough. Perhaps it is helpful to appreciate that we are dealing with a group of individuals who are often only partly competent. This is what is "wrong" with individuals with MR/ID. The textbooks refer to lower IQ's, lower adaptive skills, etc. This is a way of saying their competencies are limited. A given individual might be competent enough only to pick what he wants for breakfast. Another might be competent enough to get married and rear a child. There are few easy answers to many of these dilemmas, but it is also helpful to appreciate that, issues of legality aside, most of us, at least some of the time, are about two parts competent and eight parts Aunt Clara.

So Just Remember This

> ➢ Historically, individuals with developmental disabilities have had their rights restricted.

> ➢ They were often placed in institutions, which provided poor and sometimes abusive care.

> ➢ We have been moving simultaneously towards providing more humane care and towards extending more independence to these individuals— sometimes these goals are in contradiction.

> ➢ The emphasis on humane care has some paternalistic elements.

> ➢ Extending more opportunity has resulted in some good things like a movement away from institutions and towards such things as group homes, apartment living, and even individual home ownership.

➢ This independence is not always an unmixed blessing because it can result in risks the person is not yet ready to handle.

➢ Whenever possible, it is a good idea to extend and increase the competencies, including legal ones, of individuals with developmental disabilities. We should remain aware of the possibilities for failure that go with this and not pursue it blindly.

➢ Even traditional institutions have a place on the continuum of services, and sometimes some distinct advantages.

➢ Sometimes we have to be willing to recognize that a particular freedom is having destructive consequences for the individual and others and be willing to step in and take it away.

➢ Individuals with MR/ID often have only partial competency, and it is difficult to decide when to step in and when to step out.

CHAPTER SIX

CARROTS AND STICKS

Over the years, I have developed the habit of referring to behavior modification programs as carrots and sticks. It has become such a habit that I can't stop and probably won't stop because change demands energy and I was born lazy and then got worse. So it's carrots and sticks. But please keep in mind we are talking metaphors here. I don't mean real sticks. I don't mean hitting people with any kind of stick to change their behavior. I suppose it would be okay to give someone a carrot as a reward if you could find someone who actually liked carrots, but I am not personally acquainted with anyone like this.

There is a solid science of carrots and sticks, stemming mostly from the work of a psychologist named Skinner. Skinner was concerned with such matters as positive and negative reinforcement and punishment, and there is a huge body of scientific research supporting the validity of Skinner's ideas. Scientists, in describing experiments, like to use the word elegant to refer to research that is so well done that it allows for only one possible conclusion. Skinner's thinking about behavior, and the research that supports his ideas, is properly referred to as elegant. Perhaps because this work is so elegant, so bullet-proof, I cannot resist the temptation to refer to it as carrots and sticks.

The scientific foundation of reinforcement theory is much more solid than that for psychotropic medicines. Many of Skinner's original experiments were carried out with animals like rats and pigeons. The same, of course, is generally true for initial experiments regarding medicines. Later, Skinner and the followers of his principles applied these ideas to a variety of human problems and described some remarkable successes. This is widely known, and we certainly employ a great many behavior modification programs with individuals with MR/ID. So if everything is so perfect, why isn't everything so perfect? The principles are sound, but we look around and see we have not been able to use them to solve all the problems experienced by individuals with developmental disabilities.

An Offer I Can't Refuse

Consider, for example, that you want to devise a behavior modification program to induce me to quit using the phrase carrots and sticks. I can think of a couple of ways of achieving this goal. One thing you could do is to offer me a million dollars never to say carrots and sticks again. Another way would be to follow me around everywhere with a gun and promise me that you are going to shoot me if I ever use this phrase again. I promise you that either method would work; "you know what" would never again pass my lips. Murder and a million dollars are pretty powerful consequences, but we typically do not have such powerful incentives at our disposal. So, the principles are doubtless sound, but sometimes we fail in practice because we either lack the million bucks to give away or there are certain ethical objections to behaving like the Godfather.

Celebrating Our Lack of Control

Fortunately we lack complete control over the individual's environment. We often get frustrated when a behavior modification plan fails; the frustration is understandable, but perhaps we should also take a moment to celebrate. We would not have failed if we had complete power over the individual's life. In general, it is more important to limit the power of behavior modifiers than always to be able to fix the problem. The principles work, but happily we do not choose to use them in such a way that they will always be effective because this would mean achieving complete control over the individual's environment; this is the same as a prison, a police state, or, at best, a benevolent dictatorship.

Since we do not want to be dictators (unless, of course, I happen to be the Big Chief), then we are always operating with reinforcement programs "at the margins." By this, I mean that we are using those few reinforcers that are available to us after the basic rights of the individual have been ensured. We cannot, for example, use food as a reinforcer although under some circumstances we might use "extra" treats in this manner. The positive reinforcers that we employ generally have to be "extras" and usually the program needs to be negotiated with the individual himself (depending upon his level of competence). Mr. Jones greets strangers on the street with bear hugs. This leaves these strangers with spines that are in much better alignment; unfortunately, they are also scared witless. We can give Mr. Jones practice in saying hi and shaking hands and give him a token every time he does this successfully with a stranger. We can then let him trade in his tokens every week for an extra trip to the movies (ideally, these movies would star someone other than Arnold Schwarzenegger). We cannot make dinner contingent upon shaking hands, and it is doubtful whether we should make regular outings contingent on shaking hands.

A recurring theme in this book is the idea of balancing acts, and this is another one. If there is some danger in Mr. Jones going on his regular outings, then we might act to limit these. For example, Mr. Jones might be in danger of legal prosecution for some of his actions, or he might, because of his strength, actually injure an unsuspecting citizen. In such cases, we might, through a joint decision of the team, act to restrict his community access and set up a behavior modification program whose underlying principle is that you have to learn to act different before you get the same outing that everyone else gets.

Punishment

Punishment programs do not see a lot of use these days, and this is probably a good thing. In punishment, we immediately follow some undesirable behavior with some unhappy consequence. I note in passing that, strictly speaking, punishment is not the same thing as negative reinforcement. These terms are often confused. Under most circumstances, it is not that important to even know what negative reinforcement is. Happily, this lets me off the hook in trying to explain it.

It is important to know what punishment is because mild punishment programs are not uncommon. Most of us are familiar with punishment programs in the context of rearing children. A child gets in a fight and then gets a spanking. Or he gets sent to his room or loses TV privileges. We hope through these devices to teach the child a lesson. There is certainly a place for some procedures of this sort in rearing children.

However, with adults with MR/ID, the usefulness of punishment procedures is generally limited. Recognizing this, most workers in the field have gradually changed their practices in the direction of using fewer of these programs and in using very mild consequences. It used to be the case that harsh consequences were employed. This included at times electrical shocks for misbehavior. It included routine use of isolation rooms (bare, small cells) for time out. Physical restraint sometimes served as a punishment procedure. Sometimes there is a place for physical restraint when the individual presents an imminent danger to himself or others, and there is no other way of heading off this danger. However, routine use of these procedures sometimes amounted to nothing more than a fight, and sometimes staff were hired because they shared certain characteristics with Hulk Hogan. They were enforcers.

The Punitive Trap

Knowing the history of these abusive practices has made us wary of punishment programs. Beware also of the mind-set that says we need to teach this individual a lesson concerning his misbehavior. This mind-set reaches out on occasion and grabs all but the saints among us. Mostly it just reflects our own anger and doesn't accomplish beans for the client. Sometimes we can't teach an individual a particular lesson because he is literally too mentally impaired to learn that lesson. Abstract lessons like the Golden Rule are simply too difficult. The same applies to trying to teach me quantum physics. You could easily waste the best years of your life trying to "teach me a lesson" regarding quantum physics. The whole enterprise would just be a recipe for mutual misery.

This is not to say that individuals with MR/ID cannot learn not to hit others or not to destroy property, etc. They can certainly learn this, but they may or may not ever connect it with an abstract concept like the Golden Rule. The best use of this abstract concept is for us to apply it to ourselves whenever we get the urge to be punitive. Had some of my own teachers, in my more difficult graduate school courses, given in to their punitive urges on all the occasions when I "didn't get it," I would be speaking to you now from the safe confines of a full body cast.

Punishment—Lighten Up

Punishment procedures, at least in their milder forms, can sometimes be helpful. One reason punishment should be mild is to keep from upsetting people. If punishments are too strong, the learner gets agitated, and agitated people don't learn much. Mild punishments are sometimes effective in helping people learn, through repetitive practice, not to do something. A common criticism of punishment procedures is that all they do is suppress responses. I have never been terribly discouraged by this criticism. I am keenly aware that I personally have quite an extensive list of responses that are best suppressed. My mother certainly devoted great energy to suppressing as many of my annoying habits as possible when I was a child, but there was only one of her set against an overwhelming tide of annoyance. Learning to suppress certain responses in certain situations is the learning of limits. This is good stuff to know. For example, I know a shaggy dog story that I think is hilarious and that illustrates the setting of limits. This story takes about ten minutes in the telling and ends with one of the worst puns imaginable. I am not going to tell this story. I am going to suppress this response. My mother would be proud.

Response Cost

When I refer to mild punishment procedures for individuals with developmental disabilities, I am having reference to such behavior modification tactics as response cost. Ms. Harrison likes to spit on co-workers in her prevocational program. Mostly she is not mad when she does this. My guess is boredom. When she thinks her day needs a little more life to it, a little more oomph, she walks around the room, finds someone concentrating on their work, and spits in their face. This makes for some pretty lively interactions. Another thing we know about Ms. Harrison is that she likes to string beads to make necklaces. So we start the day with an extra supply of beads in a jar at the front desk. We explain to Ms. Harrison that the beads are hers at the end of the day if she refrains from spitting on anyone. Every time she spits, we say, "Oh, too bad," in a calm voice and take one bead from Ms. Harrison's jar. This is a punishment program called response cost. The punishments are very mild. One small punishment can be administered every time Ms. Harrison fails to suppress her impulse to spit. We design the program in such a way that everyone stays calm. No one makes a big fuss as the punishments are administered. We don't try to teach her a big lesson all in one day; rather, we rely on repeated practice sessions over time.

Staying Positive

I mentioned in the above paragraph that perhaps Ms. Harrison spits on people because of boredom. This notion takes us into the realm of positive programming. Most behavior modification programs these days are of this sort. Often these programs involve actually giving the individual reinforcement for engaging in some desirable action. More generally, however, we might just focus on providing the individual with positive activities that are reinforcing all by themselves. You don't have to reward me for going skiing— the activity is intrinsically reinforcing. So, if Ms. Harrison is bored at work, we might scratch our heads and look around for a more interesting job for her. It is possible that this step, by itself, would be enough to get her to stop spitting.

If this did not work with Ms. Harrison, then we might give her an external positive reinforcement for nicer interactions with others. We could give her some reinforcement for shaking hands with others. Ideally, we would like to think of some response to reinforce that is incompatible with spitting; it would be nice if we could be clever enough to think of an alternative response to reinforce such that if you are doing the one thing then by definition you can't be doing the other thing. This is hard to do for spitting— reinforcing swallowing is not very practical. I do recall an instance in which it helped to reinforce the wearing of tennis shoes rather than cowboy boots because it was preferable to be kicked by a softer shoe. Sometimes the best we can do is rely on the fact that no one is likely to want to shake Ms. Harrison's hand after they have just been spat upon. It is important to keep sequencing in mind as we reinforce Ms. Harrison for shaking hands; we would prefer that Ms. Harrison not learn a sequence in which first you spit, then you shake the hand of your victim, and then you get a reward.

Weak External Reinforcement vs. Powerful Intrinsic Reinforcement

This spitting example takes us back to the example at the outset of this chapter. I mentioned that you could change my behavior quite easily by giving me a million dollars or shooting me. This is a powerful positive reinforcer and a powerful punishment. Often we do not have such powerful events at our command. Giving someone extra beads to string is not quite in the same category of powerful events. We are often left in the situation of having pretty weak external reinforcers, and with these weak external reinforcers we are trying to influence a behavior that has powerful intrinsic reinforcement. Perhaps you were to decide that my skiing was a bad habit that you wanted to get rid of. You had best offer me something a little more powerful than beads if you expect to get me to stop engaging in this activity I find so satisfying.

Consider, for example, aggression. Aggression is often its own reward. People continue with hitting or with bullying because it works. If I can threaten you and make you give up your soft drink, then why in the world would I want to stop? If I can fake a seizure and get attention for this, why would I want to stop? If I can spit on other clients and staff, the reaction I get may be a lot more satisfying than a couple of beads. Native Americans gave up Manhattan Island for this bead scam. No, thank you very much, I'll stick to spitting. Staff can't hit back or spit back. Once in a while, another client may whomp me one, but this happens on such an irregular basis that it is no real deterrent.

Sometimes the behavior modification problems are easy ones. This is often the case if someone is doing something annoying for attention. If we establish that attention is the reinforcer, it should certainly be fairly straightforward to give the individual attention when he is doing something constructive; this allows us to promote a constructive activity and satisfy his need for attention at the same time. On principle, the annoying behavior should reduce in frequency. And we should all live happily ever after. Sometimes things really are this simple.

Crowding Out Unwanted Behavior

At other times, things are not so simple, especially when the individual has a life-long pattern of such intrinsically reinforcing behavior as hitting or property destruction. In this circumstance, sometimes it helps to simply crowd out the destructive activity with more productive activities. We can go to special pains to fill the individual's time with intrinsically satisfying stuff like shooting hoops or doing crafts, etc. This approach is often especially useful when the troublesome behavior occurs at a particular time of day. For example, the individual often gets in a fight in the first hour or two after returning home from work. If we fill this time with basketball, we are reinforcing an alternative behavior, and our job is made doubly simple by the fact that this alternative behavior may require no external reinforcement. Basketball, like skiing, can be intrinsically reinforcing.

First the Peas, Then the Ice Cream

In other situations, we can use an intrinsically reinforcing activity to get the individual to do something he would, all things considered, rather not do. All parents use this principle when they make dessert contingent on first eating disgusting and revolting vegetables. In this case, first we do household chores, and then we shoot hoops. Or, first we sit down together and resolve roommate conflicts without recourse to violence, and then we shoot hoops.

This chapter is not intended to make anyone an expert on behavior modification practices. For those wishing to learn more, there are a zillion good books on the subject. I especially recommend the treatments provided in the books by Lavigna and Thompson; these books place special emphasis on positive programming. I note that there are many behavior modification programs that require a great deal of technical expertise to devise and administer. Again, this takes us back to the dilemma we discussed earlier in the chapter—we may know how to achieve control over some behavior through technically sophisticated programs, but as complexity increases often so does the need for control over the individual's environment. This can make the program difficult and expensive to administer and often raises ethical questions concerning the rights of the individual who is being "modified."

A Magic Trick

One of the most fascinating behavior modification procedures involves the differential reinforcement of low rates of behavior (DRL). Magicians who saw their assistant in half typically warn the audience not to try this at home. Similarly, DRL's require training and judgment to devise and administer, so don't try this unassisted. The basic idea, however, is fiendishly clever. We count the frequency of some undesirable behavior like spitting. We find the individual spits on someone eight times an hour on average. We then give the individual an external reinforcer for spitting eight times an hour (but not nine times). We are reinforcing the individual for engaging in the undesirable behavior. Once we have done this for a period of time, we change the economy. Now you only get the reinforcement if you spit seven times an hour. We continue to drive down the number of spits through reinforcing progressively lower rates of behavior. I still have trouble believing this actually works but, done right, it can.

Start at the Front

Another notion that can be employed in behavior modification programs is to work on the front side of the behavior rather than the back side. Reinforcers and punishers follow behavior. It also helps to look at what comes on the front side of the behavior. What is the stimulus that sets the stage for the behavior? Ms. Chambers hits people when the room gets too noisy. After her vocational program, she waits on transportation in the commons area of the building along with about 20 other people. It's noisy, and Ms. Chambers gets upset and hits somebody. We don't need to reinforce her for not hitting, or reinforce her for low rates of behavior, or reinforce her for more appropriate ways of problem-solving, or give her a mild punishment every time she hits. We need to buy her some headphones and let her shut out the noise with soothing music. Or let her wait somewhere else. Sometimes it's a lot easier to work on the front side of the problem behavior than the back side.

Count Carefully

A wonderful side-effect of behavior modification programs is that they require us to pay close attention to exactly what is going on in the person's life. In the example above, we should certainly start by asking Ms. Chambers why she gets upset and hits people. However, sometimes she can't tell us or simply does not know herself. We have to make careful and systematic observations to determine whether it is the noise that is setting her off or the number of people or a particular person she does not like or whether others are having snacks and Ms. Chambers is not, etc. Behavior modifiers refer to this careful observation as a functional analysis of behavior. Behavior modifiers not only push us to observe carefully but also to count. We need to count and keep careful records so we can grab some clues about whether we are actually making progress. Counting also helps us to make accurate judgments about problems. Sometimes a staff member is having a bad week and begins filing reports that Mr. Norton is being more oppositional and noncompliant. Without defining the behavior and doing some counting, it is difficult to know what the problem is, whether Mr. Norton is actually doing more of it than usual, whether the staff member simply needs a couple of days off, or whether Mr. Norton is acting like a jerk because he accurately perceives that his staff person is acting like a jerk, and Mr. Norton has decided not to put up with any more of this crap.

Simple and Immediate

Behavior modification programs need to be as simple and easy as possible so staff can administer them effectively. Most people are familiar with the rule that consequences need to be immediate. We often use tokens like poker chips in order to bridge the time gap between the behavior and the actual reinforcer. If the reinforcer is a weekend outing, then this by definition cannot be administered immediately for good behavior at work on Monday. Also, weekend outings or craft supplies, etc. are not nearly as portable as poker chips. If we want to reinforce Mr. Smith each time he shakes hands rather than giving bear hugs, we need to give the token immediately the handshake occurs. This can be difficult for staff because at the moment Mr. Smith is shaking someone's hand, staff are fully occupied helping Ms. Jones clean up a toileting accident. By the time they get around to giving Mr. Smith the poker chip, he has just spilled his drink on the carpet. Just like medicines, reinforcement programs can have some unintended side-effects if we are not careful.

The Joy and Heartbreak of Partial Reinforcement

All the examples up to now have assumed the use of continuous reinforcement, i.e., every response gets a reinforcer. The real world is not like this. Mostly what goes on in every day life is partial reinforcement, in which not every response gets reinforced or punished. When we are trying to teach new skills, reinforcing every response is best. Once the individual has learned a response, however, his responding is maintained best by partial reinforcement. Keep in mind that this principle applies both to desirable responses and to undesirable ones. Habits, both good and bad, are maintained through partial reinforcement. Much of Skinner's scientific work was devoted to describing how individuals behave under different schedules, different arrangements, of partial reinforcement. What happens if you reinforce every third response but only make the reinforcement available at set intervals, e.g., every hour? What happens if you reinforce every third response at intervals that average out to an hour such that sometimes you reinforce in thirty minutes and sometimes in an hour and a half? What happens if you forget the time intervals and simply reinforce every third response whenever it occurs? What happens if you reinforce on average every third response such that sometimes the individual gets a reinforcer after one response and sometimes after five responses? Students of Skinner know the answers to all these questions. Further, Skinner's disciples have extended this body of knowledge such that they can predict the effects on behavior of very complicated sequences of reinforcement. This is one of those areas where, as a general principle, the scientists have made more information available to us than we are able to effectively make use of. In actual practice, we have trouble making use of some of this knowledge because, as mentioned above, we lack sufficient control over the individual's environment. Or we lack the staff and resources to implement complicated systems.

Although we cannot always use this knowledge to fix things, it does sometimes help us to understand things. Sometimes people develop habits that seem to be self-defeating and baffling. We cannot, for the life of us, figure out why they keep doing it. If we know enough about the individual's history, we can usually discover that at least part of the answer lies in the power of partial reinforcement. There has usually been a history of some sort of pay-off part of the time, and this is often enough to keep the individual going almost indefinitely. Over the years I have been a dedicated fly-fisherman. If you were hired to be my attending staff, you might drop in on me one day while I am in the river fishing. You could then observe an incredibly high rate of casting behavior. Depending on when you dropped in, you might observe that this behavior results in no reinforcement. As you sit there on the bank of the river, you ask yourself why this pathetic creature keeps casting the fly upstream and letting it drift down, casting upstream, drifting down, upstream, downstream. This pathetic creature is in the powerful grip of a partial reinforcement schedule because once in a while he manages to fool some especially gullible trout.

The Dreaded Extinction Spike

A related phenomenon is referred to as the extinction spike. One way to make undesirable responses go away is simply to withhold all reinforcement for this response forever. The only difficulty with this approach is that this simply isn't so simple. If the individual is engaging in repetitive tapping for attention, then it would seem that we could simply withhold attention and the individual would eventually cease the tapping. In principle, this is true, but in practice it rarely works. The first thing that happens when we try this is the extinction spike. When the individual does not get what he expects from the tapping, he taps harder, he taps louder, he taps longer. So the immediate result of this procedure is to increase the problem.

This is usually about the time that we run afoul of the heartbreak of partial reinforcement. Another staff member drops in, and this staff member has not been clued into the plan. Immediately this person begins to give the client attention for his tapping, and the client has achieved his goal. From this exercise, the client learns that when he can't get attention for being bad, his best course of action is to be really, really bad. The moral to this story, of course, is to use some version of positive programming, i.e., give the individual loads of attention for behaving positively and make positively sure you are watching like a hawk to catch the individual being good.

It's About Life

Behavior modification programs should not be considered as anything separate from the more general task of improving the quality of the individual's life. Sometimes they are necessary and useful. But it is important to put them in the context of more general issues involving problems in living and interpersonal relationships. Behavior modification programs are not always necessary. Believe it or not, sometimes people change their behavior if we just ask them nicely. This radical approach is sometimes overlooked in our rush to create a program. If I have a roommate I despise, no one creates for me any programming, positive or otherwise, to help me get along better with my roommate. I would simply get another roommate or live by myself. There are times we need to stuff the program in a drawer or some other appropriate receptacle and help the individual find another roommate, another job, etc. Behavior modification, like psychotropic medicines, can be a useful tool; regrettably, neither tool spares us the necessity of careful thought about costs and benefits.

So Just Remember This

➢ Behavior modification programs refer to the systematic use of reinforcement and punishment to change behavior.

➢ The principles are sound and rest on a solid scientific foundation; however, they are not a magic answer because we often lack sufficient control over the person's environment.

➢ The fact that we lack sufficient control can be a good thing—remember freedom?

➢ We have to use the (sometimes limited) reinforcers that are available to us after the basic rights of the individual are ensured.

➢ Punishment programs are those that immediately follow some undesirable behavior with a consequence that "suppresses" the behavior.

➢ These programs are not used frequently. When they are used, they tend to work best if the consequence is very mild, as in a response cost program.

➤ It is important to avoid a punitive mindset.

➤ Positive reinforcement programs are much more widely used. These can involve external rewards or the opportunity to engage in activities that are intrinsically reinforcing, like sports.

➤ A difficulty arises when we are trying to influence a behavior that has powerful intrinsic reinforcement with a weak external reinforcement.

➤ In devising programs, we often have to balance complicated ethical questions that arise when we try to achieve a positive outcome by doing things that may reduce a person's freedom.

➤ A powerful procedure in behavior modification is to change not the consequence but the thing that comes before or sets off the behavior.

➤ A good thing about behavior modification is that it requires us to make careful observations and count behavior—a functional analysis. Sometimes a good answer leaps out at us when we make such observations.

➤ Much of our behavior is governed not by continuous reinforcement but by the power of partial reinforcement.

➤ Behavior modification can be a useful tool but is best understood as part of the general task of improving the quality of life; as with all such things, it requires humanity, good sense, and a consideration of costs and benefits.

CHAPTER SEVEN

PAY ATTENTION, PLEASE

A nice thing about working with developmentally disabled folks is that you don't have to learn very much that is new to do it. You only have to put to good use what you already know. What could be easier? As with most things, however, there is a catch. The catch is that most of us almost never put to good use what we already know.

One thing we already know is that it helps to pay attention to people. I know this; you know this; there is almost nobody in the whole wide world who doesn't know this. But mostly we don't do it. What we do is pay the finest of lip service to the idea and go on about our business. And when we do actually give someone our pure and undivided and unadulterated attention, it is too often because they are powerful or smart or beautiful or sign our paychecks.

Developmentally disabled folks are often none of the above. They are not smart in the way this is usually defined. They are sometimes not physically pretty. Usually they are poor. They do not sign our paychecks.

But they do need attention. Like most folks, they are encouraged by positive attention. And, like most folks, if they fail to get positive attention, they will go for its opposite. Our job is to encourage people, and the most powerful form of encouragement is positive attention. Everybody knows that—everybody's grandmother knows that. We don't need high-powered (or even low-powered) shrinks to tell us that; what we need to do is implement what we already know and implement it in ways and amounts that most of us are unaccustomed to.

Hiding from Susan

I have been fortunate over the years to become somebody special to a woman with developmental disabilities who used to drive me crazy. Call her Susan. Early in our acquaintance we had a routine appointment or two to get acquainted so I could consult on some of her programs. Subsequently I often ran across her in the halls of the office building where I work. I am a busy, important person. If I am in the halls it is because I am rushing somewhere, maybe to an appointment with some other busy, important person. Maybe to deliver busy, important-person papers to my secretary. Or maybe to go to the busy, important-person john. Susan would grab me in the halls and want to talk. This was fine—I was pleased to pause for a second and say hello and inquire after her welfare. Susan, however, had no appreciation of the busy, important-person rules. She did not understand her job. Her job was to say, "I'm fine, thank you," and then to let me be on my way.

I discovered that when you are grabbed in the halls by Susan you are most definitively grabbed. And you stay grabbed until Susan decides to let go of you. And Susan only decides to let go when she can no longer think of anything to say. And Susan can always think of something to say until she runs out of breath. And Susan, through a marvel of biology not yet fully understood by science, does not need to breathe when she talks. And when you try to get away, she talks louder (still without breathing). This is one of those conversations where, when it doesn't go right, you find that Susan is screaming at you at the top of her scientifically amazing lungs, holding on to your shirt with her left hand, and hitting you all over your important-person facial area with her right.

One of the things I learned from this was to hide in my office all day and never come out. I did this for awhile, six months or so, but people were beginning to think I had become deceased. The six months was not wasted, however, because I was able to fashion all the paper clips in my office into an extremely long chain and also to come up with a solution to the Susan dilemma.

Getting Brave

I boldly ventured into the hall. Susan, who had been lying in wait outside my door for six months, grabbed me. I made no effort to escape. I looked her in the eye. When she grabbed my arm, I put my hand on top of hers and patted her hand. I listened intently for five minutes, and then I interrupted and thanked her for taking the time to talk to me and asked her if we could go make an appointment with my secretary so we could talk some more about this. She was delighted. I gave her the card with the appointment time on it. I was free.

This story has to do with the power of attention. I have scheduled a weekly appointment with Susan for the past seven years. We meet for about ten minutes and have a cup of coffee. What we do is certainly not psychotherapy as they taught it to me in school. Susan's job is to talk, and my job is to listen intently and nod and smile at the correct times. Susan does not encourage me to talk much, and she certainly would consider it a silly waste of her time to sit and listen to a bunch of unwanted advice from me. We have settled into a ritual. When I see she has finished her coffee, I ask if she would mind if we made another appointment for next week. I ask this with elaborate courtesy as if she would be doing me a favor to work me into her schedule. She always happily agrees, and the appointment is over. It has taken all of ten minutes. She no longer needs to collar me in the hall because she knows we will be having an appointment in which she can tell me what is on her mind. This has turned out to be a much more efficient use of my time than hiding in my office or being grabbed for interminable intervals in the hall.

Wising Up

There has also been another, unanticipated result. I have spent seven years acting as if she were doing me a favor by making another appointment with me. As may be clear, I am not the world's fastest learner—it took me six months to quit hiding in my office. And it took me seven years to realize that it is Susan that has done me the favor and not the other way around.

She has shared coffee and gossip and complaints and fits of rage and fits of laughter with me, and we have become part of each other's routine and pieces of each other's lives. She thinks of me as a friend. Susan is apparently tone deaf because on her birthdays she likes it when I sing Happy Birthday to her.

She has done me this wonderful favor, and all I had to do to bring it about was to shut up and pay attention to her.

We all have it in us to give the gift of our attention to others. We just don't do it much.

Not Now—I'm Busy

We are, as the popular phrase has it, task-oriented. We know we have accomplished something when we have run through the bicycle safety program or the laundry program. We have accomplished something when we get clients delivered to the doctor's office on time or prepare the grocery list. In the process of all this accomplishment, we don't manage to pay much attention to clients. Not only do we not pay attention, we run and hide like I did with Susan. We say, "Not now, I'm busy." We are frustrated at their inability to understand that we are overwhelmed with the demands of meeting their needs. Clearly we are busy meeting somebody's needs, but it is, of course, an open question as to whose needs are being met. Maybe ours. Maybe our supervisor's.

The Best Program There Is

If we insist on running some program, let it be the program called: I AM NOW GOING TO GIVE YOU MY FULL AND UNDIVIDED ATTENTION. This is a pretty good program. It gets results.

By definition, it works best when it is one on one.

It works best when it is initiated by staff. "There you are, Bill, I've been looking for you because I wanted to spend some time with you."

It works best when it is frequent. My example, above, of seeing Susan once a week, is not frequent enough. The needs of people vary. But some people need attention once a day or three times a day or once an hour or once every thirty minutes. The frequency may vary, but it is important to keep in mind that rarely is anyone seriously injured by an excess of attention.

It can be instructive to simply spend several hours sitting and watching the interactions in a group home or similar facility. Clients approach staff with requests or complaints. Staff approach clients with reminders to do chores or take medicines or come to dinner, etc. Staff, like me, do not often enough approach clients just to talk or shoot hoops or play a board game. When asked about this later, staff usually respond by saying that they give clients lots of attention.

Attention needs to be given intensively and on a much more frequent schedule than most of us are accustomed to. Consider approaching clients every half-hour and spending from two to five minutes with that individual client. Try it as an experiment.

Some Practical Tips

In that five minute interval, listen. Make eye contact. Touch the person's hand or elbow. Ask how they feel. Use reflective listening techniques, which are described further elsewhere. Explore the client's feelings about events that have occurred that day. Use visual aides, if necessary, to picture activities or describe events. A nonverbal client can describe his day at work by sorting through pictures of himself as he engages in various work activities. Preferred activities can also be pictured in the same way. Emotions can be discussed with the aide of drawings of happy faces, sad faces, etc., or, even better, with pictures of the client himself when he is sad or mad or happy. A Polaroid is a useful tool. Similarly, it can help to have a picture book of important people in that person's life (relatives, favorite friends or staff) to use as a basis for conversation.

Identify a transition device with which to end these sessions. This might be a token to give to the client that is redeemable for individual time during the next hour.

Remember, this is not a punishment. If the client has something else he wants to do, he is not required to sit and have a long talk with staff. But the idea here is to try to make the time together fun and rewarding so that most of the time the client will want to be involved.

Something You Can Take Home

Once staff members have tried these approaches they often find the same thing helpful at home. I am of the opinion that you cannot give children too much love or too much attention. Most of us end up talking with our children during the TV commercials or on the way to school or at other odd moments as they arise.

Do not wait for the odd moments. Schedule time with your children. Schedule time with your spouse. Most of us complain that we cannot afford the time to do this. The fact is that most of us cannot afford not to schedule this time.

I spend half my time working with individuals with developmental disabilities and the other half in the private practice of clinical psychology. Much of this private practice work is with children, so I often find myself giving parenting advice. I routinely recommend to parents that they set aside each night five minutes of "special time" with their child. This is, by definition, individual time so that if they have more than one child, then I recommend that they set up separate sessions with each child. In making these recommendations to parents, I typically explain the logic along the lines described above.

The reactions I get from parents are instructive. Once in a while a parent explains that this is too much time and that they are too busy; this response, however, is rare. More typically, parents are skeptical because they do not think such a small amount of time could have an impact on the serious problems their children have. At the same time, however, when I probe for their daily routine with the child, it is apparent they are not spending even this amount of time in giving the child the kind of intensive attention described above. I am left scratching my head at this puzzlement. Often they seem to be saying that they can't do an hour of individual attention because it is too much and won't do five minutes because it is too little.

I think some of their resistance stems from a desire for complicated and mysterious answers rather than simple ones. Perhaps they feel they are not getting their money's worth when they pay me a hefty fee to tell them something they already know. We are back to our starting point. Not only am I telling them something they already know, I am telling them something everyone already knows.

I expect it is a bit aggravating. There is a wonderful story about a famous and expensive attorney in Washington, D.C. A large corporation was not satisfied with the wisdom of their in-house attorneys and sought the counsel of this famous man. He listened to their problem and spent the afternoon reading the documents they brought him. The next day he wrote them a letter that consisted, in its entirety, of the advice: "Do nothing." Attached to this rather brief letter was a bill for a half a million dollars. The president of the corporation called to protest, whereupon the famous lawyer explained that the right advice was worth the price.

It's the Simple Stuff that Holds Us Together

The advice does not have to be complicated to be valuable.

After more years of training and experience than I can cheerfully recall, I have no grand theory as to why we need attention.

The question of why we need attention is not unlike the old philosophical puzzle about the tree falling in the forest when there is no one there to hear it. I am no expert at philosophical puzzles, but I like to be heard on occasion, to be noticed on occasion, and if I crash in the forest, or even in the living room, I like to have it remarked upon. Absent gravity, the planets would drift alone in space, each pursuing its own odd path. Attention, and the caring inherent in it, provides much of the force that bends our orbits, that binds us, each to the other, and keeps us from being set adrift.

So Just Remember This

> ➤ **The power of positive attention in working with individuals with developmental disabilities cannot**

be over-emphasized. This is something we all know how to do—we just don't do it enough.

➤ There is no more powerful "program" and no more important task than giving someone your full and undivided attention.

➤ Try giving clients one on one attention. Be the one that initiates the interaction. Do it frequently, far more frequently than you would ordinarily think necessary.

➤ Use whatever clever ideas and adaptive tricks you can think of to promote conversation. It works, with clients, with family, with almost everybody.

CHAPTER EIGHT

KING OF THE HILL

There are often power struggles in working with individuals with MR/ID. As with most things, there is no answer to these problems that is universally right. Sometimes it helps, however, to carefully examine how we think about these dilemmas.

Whenever you find yourself in a power struggle, consider a radical alternative: consider giving up. Don't just give up after a struggle. Give up quickly. Find a white flag. Wave it. Give up and give the individual what he or she wants. We often tear our hair out when working with folks to get them to communicate with us. When we are in the midst of a power struggle, it is very clear that we have succeeded at one thing. We have received a communication. We have learned an important thing about the individual. They are showing us something they want. We are supposed to be there to serve. So give them what they want.

Okay. Okay. Not in all situations. If Mr. Smith wants to drink Drano with dinner, then it's time to have a power struggle. If Mr. Smith wants to serve Drano to someone else for dinner, then it's time to have a power struggle. But most situations are not of the dinner and Drano variety. More often, Mr. Smith is sitting down to dinner in his own private home with five other developmentally disabled folks he did not pick as roommates and a staff member he did not hire.

I don't live like this. I don't share my home with six people of somebody else's choosing. If I did, I can almost guarantee it would make me cranky. At my home, I have something to do with picking the menu. If I didn't, I would be cranky, maybe even a notch or two past cranky. I can also promise that if I had to share my home with six people I didn't pick, I would figure out a way to despise at least five of them.

So Mr. Smith is sitting down to dinner in what is marginally his own home to a meal whose menu is not of his choosing with a bunch of people he would sooner leave than take. And, wonder of wonders, there is a bowl of new potatoes on the table in front of him, and he loves new potatoes. So he reaches over and takes the serving spoon and carefully serves himself all the new potatoes. Those that won't fit on his plate he neatly stacks on the table in front of his plate. Mr. Smith has a serious interest in new potatoes.

Mr. Smith's approach defies everything we know about fairness, good manners, sharing, and the American way. Staff are likely to feel responsible for defending to the death all of the above values so they rush in and try to wrestle away the new potatoes. This is a power struggle. Give yourself the Drano test before rushing in: Is this Drano or new potatoes? If it flunks the Drano test, then let Mr. Smith have the potatoes.

King of the Hill

There are times, in August, when I have been known to make a meal on fresh sweet corn and peaches. I fix what I want to eat; I eat as much of it as I want, and I am not pestered about sharing and the American way.

If Mr. Smith can be encouraged to share, wonderful. If not, make some more new potatoes.

Power to the People

Power struggles are designed to take control from individuals, and they are hard to resist. As soon as you have finished making more new potatoes, the first impulse that a person has is usually to sit down and start looking for ways to teach Mr. Smith to cooperate and share and be a real American. My suggestion is to sit down and start looking for a way to give Mr. Smith more control. Can he have more control over menu selection? Can he have more control over mealtime? Can he have more control over where he sits? Over what he wears? Over what TV station he watches?

When there is a power struggle, we should be looking first at ways to give up control. The client may benefit from having more control and may become an easier person to get along with. In some situations it may also be the client's right to have control and make certain choices, and it makes not even a little bit of difference whether he becomes a nicer person or not. Sometimes if we avoid the power struggle, this gives the client time to calm down. I have noticed about myself that I am easier to deal with and much more agreeable to compromise when I am not in the midst of a blind, murderous rage.

Help People Notice Their Choices

Another way to provide the individual with more control is to emphasize to the individual the choices he already has. It often helps to repeatedly remind individuals of their choices throughout the course of the day. This can be worked into conversations on a natural basis.

"John, do you want to wear your sneakers or your sandals?"

"Bill, I see you picked your red shirt to wear. I think you must like that shirt."

"Mary, do you want to sit in the chair or on the sofa to watch TV?"

"Harry, I see you decided to shoot some hoops, You're really into basketball."

Part of the logic here is that you want to call attention to the many choices each individual makes throughout the day. We all often need to be reminded of how many choices we have as part of our daily routine. These reminders can promote feelings of control and add to the satisfaction the individual feels with his day.

Years ago I used to teach on occasion a course in abnormal psychology to a group of inmates in a college program in the Texas prison system. This was always an interesting experience, especially when we got to the chapter on antisocial personality disorders.

More to the current point, it sometimes comes back to me very vividly the feeling I had each night when I left the prison. When the bars clanked shut behind me, I could walk left or walk right or sit on the grass and do nothing or go out to dinner or almost anything I pleased as long as it was not sufficiently illegal to get me put back in the place I had just left. Most of us, mentally retarded or not, spend most of our time blind to our choices. We forget that even our routines are the product of our choices. It helps to find natural ways to remind clients of the control they have over many aspects of their lives.

Don't Sweat the Small Stuff

Most experienced supervisors could quickly draw up a list of things not to get into power struggles over. Here is a sample list that took me all of ten minutes to gather by stopping a couple of supervisors in the hall:

1. Ms. Jones wants to wear her plaid shirt with the polka dot pants today. Let her.

2. Mr. Smith wants to wear his Bronco's cap for the ten-thousandth consecutive day. Let him.

3. Ms. Jones moos like a cow as she pours milk over her breakfast cereal and laughs hysterically at her own joke. Actually I find this to be a pretty good piece of humor, but I have always been a sucker for cow jokes. I personally favor the one that ponders why

it is that cows have to wear bells. The answer, of course, is that their horns don't work. If Ms. Jones likes her cow joke, let her have it.

Giving up on power struggles can make for happier clients. It can also make for happier staff. Giving up some of the need to control can help staff relax. Staff sometimes create pressures for themselves by identifying their own success with the behavior of the clients with whom they work. Staff should not fall into the trap of identifying their own success with whether or not their clients are "good." Staff are responsible for following certain rules and regulations. Staff are responsible for being encouraging to clients. They are not responsible for creating perfect clients. To think otherwise is a pretty fair recipe for misery.

Clients sometimes progress very slowly. Under these circumstances, a little patience, a little faith, and a sense of the long view can be helpful.

Look for Other Power Struggles to Let Go Of

As described above, power struggles with clients can be emotionally draining and are often counterproductive. As bad as these are, however, they are as nothing compared to power struggles with supervisors. I have been a supervisor. I can tell you that I have spent years developing deeply ingrained maladaptive behavior. As a supervisor, I learn slowly. I fly into rages for no reason and am usually obsessed with paperwork and forms. In addition, of course, I have the emotional maturity of a three-year-old. It is fruitless to get into power struggles with individuals like me. It is usually best to just try to figure out what I want and give it to me.

Once a person begins trying to figure out how to avoid power struggles rather than how to seek them out, then sometimes these tactics can be employed at home. Spouses, children—there is always someone out there looking for conflict. Keep the Drano test in mind. Let people get in the habit of controlling themselves rather than needing to be controlled by you. Most of us, most of the time, try to control most things, fail miserably and make ourselves miserable in the process. I am not personally acquainted with very many exceptionally wise people because I find that we don't have much in common. However, I have noticed that the few wise people I do know spend more of their time focused on controlling themselves rather than trying to control others.

So Just Remember This

➢ **The first option to consider when in a power struggle is surrender.**

➢ **Most of the things we struggle over are not worth the damage to the relationship that a power struggle creates.**

➢ **Ask yourself if this issue is crucial—does it involve truly dangerous or damaging behavior?**

➢ Pay attention to what the other individual is trying to communicate by being willing to struggle for control.

➢ Find ways to extend to others as much control as possible over their lives.

➢ Find ways to remind others of the many choices they already do exercise control over.

➢ Remember that individuals with developmental disabilities may feel that they live in a world with less control than they would like.

➢ Remind yourself that you are not personally responsible for everything—this makes it easier to give up your own need for control.

CHAPTER NINE

SAY IT SIMPLY

Over the years I have developed a minor reputation as someone who can communicate well with individuals with MR/ID. As nearly as I can puzzle it out, the view seems to be that I have some special knowledge of individuals with developmental disabilities that flows from my training and experience. Hence, I can communicate where others fail, can lift the veil and peer into the snarled recesses of this person's mind, and can patiently pick apart the mental knots through the sheer power of the well-chosen word.

This is pretty much baloney.

Lose the Big Words

I do, however, have a secret or two, and, as I look at those who are effective, I notice they share the same secret. They speak simply. They eschew complexity in symbolic interchange—as I said, they speak simply.

This is hard to do. But it is a good thing to do. Once in a while, it helps to remind ourselves of the basics. People with MR/ID do not know a lot of big words.

Big Words Don't Make the Ideas Bigger

It is hard to learn to speak simply because we live in a society that is plagued by fancy language. It is a society that confuses brain size with word size. We can buy big cars to show off the size of our bank accounts, and we can use big words to show off the size of our brains. We can also use big words to puff up small ideas into ones that seem big and important—Cocoa Puffs for the brain.

Think about the term interdisciplinary team. If I am the client, what possible meaning could this have for me? Am I in trouble? Is this a team in charge of discipline? Will it be the Pittsburgh Steelers? I begin to wonder what it is that I have done so wrong. I imagine all sorts of punishments. And the worst of these punishments is that I may have to go to a meeting and sit around a large table and have a bunch of people beat me about the head and shoulders with big words.

They may tell me I have trouble with "transitions." If I have been really bad, they may talk about my "functioning level" or my "sub domains of adaptive functioning." They may ask me if I have "performance anxiety" or "unresolved grief issues." They may speak of my problems with "articulation." If I am really in trouble, they may threaten me with a "dialogue" or even, God forbid, an "interface."

It is bad enough that mental health workers beat each other up with such language. It is practically criminal when we use it to make miserable the lives of our clients. I have a bit of a hearing problem.

Sometimes I ask people to speak up, and they don't. They go right on mumbling. I get too embarrassed to keep asking them to speak up, so I start smiling and nodding and making guesses about what it is they might have said. I once left an administrative meeting thinking the consensus of the group had been to muster the ducks at dawn. I smiled and nodded when people turned to me for my opinion, but I did leave with the vague feeling that there might have been something I misunderstood. I expect that our clients leave most meetings with this vague feeling, not because they have a hearing impairment, but because that have been "jargoned" into submission.

I marvel at the patience our clients show us. I can recall sitting in a staffing where there was a client named Bill with moderate MR/ID and half a dozen staff who had had Cocoa Puffs for breakfast. Each staff member seemed to try to outdo the last with a plethora (sorry, I couldn't resist) of big words. Bill smiled. Bill nodded. Finally, the case manager turned to Bill and asked if he "understood the implications of our discussion." Bill smiled. Bill nodded. We all signed our names to a piece of paper.

After the meeting, Bill stopped me in the hall, and we went to my office for a cup of coffee. Bill asked for cream. We agreed it was good coffee. We both wished it would stop snowing because we were tired of shoveling snow. Bill pointed out his new snow boots, and I admired them. After a polite interval of small talk, Bill asked me what the meeting had been about. I told him I was not sure, but I thought it might have had to do with mustering the ducks at dawn. Or perhaps on the lawn—I told him I was not exactly sure about the last part, but I was pretty sure something had to be done about the ducks. He shrugged philosophically and agreed I was probably right. We both allowed as how the coffee was good and that the snow was pretty, but we hoped it would stop falling soon.

Slow Down

A nice idea for effective communication with individuals with developmental disabilities is to let conversations develop slowly. On average, individuals with MR/ID do not think as fast as others. The technical catchphrase here is that they often need more time to process information. Another reason the conversation needs to be allowed to develop slowly is that many individuals with developmental disabilities have various speech difficulties. Articulation problems often do not make for a snappy conversation.

I grew up in the Deep South, so slow-moving conversations come naturally to me. Many staffers, however, have great difficulty with slowing themselves down enough to make the conversation comfortable for their clients.

Keepinmindthepaceofiterlocutionthatdifferentindividualsfindcon genialvarieswidelyandyouwouldnotlikeitifyourinterlocutorspokesora pidlythatitseemedallthewordsrantogetherandweretoyounonsensicalb ecauseoftheirrapidity.

If figuring out the above sentence gives you a headache, then remember that headache the next time you have a conversation with someone who does not think as fast as you do. And slow down. And put in lots of pauses. Allow for some silences. There is nothing wrong with a little silence mixed in with a conversation; real life is not like TV—a little dead air can be tolerated without the world coming to an end. A nice, freshly baked loaf of bread is full of dead air—it gives it body.

Give People Time to Think

There is nothing wrong with taking time to think. Individuals with MR/ID often need this time. Often so do I. Over the years I have testified countless times as an expert witness in court. The first time I did this I was, of course, scared to death. When you testify in this manner, one of the first things you do is explain your credentials, and the judge officially declares you to be an expert. I found this to be pretty awesome, but, being scared to death, I did not think I would be able to remember a single shred of this expertise that the judge had just officially certified me as having. It came to me that when asked the first question I was going to sit there mute and frozen and that both the judge and I were going to look like idiots for thinking I was an expert in anything. My dilemma was compounded by the image I had then that an expert is supposed to know pretty much everything about pretty much anything and give quick, sharp answers, preferably etched in stone.

This is mostly not the way it worked out on that first attempt, and I still feel badly for that poor judge, who will forever bear on his record the shame of having certified me an expert when he should have certified me a babbling idiot.

Gradually I got better at being an expert witness. My progress at this developmental task was aided immensely when I finally discovered that there was no rule requiring me to blurt out quick answers. One day, in the face of a particularly vexing question from a particularly vexing attorney, I said, "That's a hard question; I need time to think about how to answer that." A blanket of unnatural silence descended over the court room. The vexing attorney was even more vexed, but mercifully speechless. I sat silently. The lawyers sat silently. The judge sat silently. I noticed the judge was scribbling notes on a legal pad. A couple weeks later, I ran into Her Honor at a social occasion and asked about her note-taking; she said it was her grocery list. During the hearing, I finally gave an answer of sorts. To this day I do not remember either the question or the answer. I do remember that I needed time to think.

Individuals with MR/ID often also need time to think. Find ways to give it to them. If you need something to do, bring a legal pad and make your grocery list during pauses in the conversation. Talk about the weather. Talk about Bill's new shoes. Begin the conversation with the familiar, the concrete. Don't ask Mary if she is depressed—ask her if she is sad. Better still, ask her if she is sad sometimes and happy sometimes because most people are not the one or the other all the time. Then ask what makes her sad. And also ask what makes her happy. It also does not hurt to ask what Mary does when she is sad to cheer herself up. I have a pretty good idea what helps to cheer me up when I am down. You probably do, also. Don't assume that Mary is any different. The best expert on someone named Mary is almost always someone named Mary.

Quality Trumps Quantity

Also don't assume that because answers may be slow in coming that they lack quality. I met the real Mary when her mother died. Mary was in her thirties. She was and is an individual with moderate MR/ID. She is thin and fragile; she is nervous a lot. She smiles when she is comfortable and enjoys teasing, but much of the time she seems to me to be lost and unsure of herself. I had met her before her mother died but did not know her very well. She has severe articulation difficulties and a thin little voice so I often have to strain both to hear and understand what she is saying. Mary had lost her father a number of years before. Staff, who referred her to me for grief counseling, told me that she had been very close to her mother.

I had my first session with her a couple of weeks after her mother died. I fixed her some coffee. We talked about the weather. Mary favors lots of costume jewelry, and I admired each and every piece—there were four rings and two necklaces. As she showed them to me, she pointed out the ones that had been given to her by her mother. I told her I was sorry about the death of her mother.

My father had died a few years before, and I told her about this and how sad I had been. I told her I was still sad sometimes when I missed my father. There was a long silence. The silence was so long that I was thinking to myself that it might take many sessions before Mary was able to talk productively about the loss of her mother.

Finally, Mary said, "My mother is dead."

There was another long silence.

"I won't be able to see her. I miss her."

More long silence.

"She is in the ground."

More silence.

"She is in heaven."

Silence.

"She is in my heart."

About half the time I find myself zigging when I should be zagging. But at least on this one occasion I had the wit to appreciate that Mary's understanding of the situation could not be improved upon. It had taken all of about ten minutes, most of which was devoted to jewelry and to silence. In the tiny bit of time remaining, Mary explained to me everything there was to know about grief, and I was, as is usually the case, more taught than teaching.

So Just Remember This

➢ Don't use big words with people who don't know a lot of big words.

➢ Most of us are not even aware of how many big words we use—pay attention to your own speech.

➢ Sometimes we need to let conversations develop slowly.

➢ Give the other person as much time to think as she needs.

➢ There is nothing wrong with a little silence.

➢ Sometimes the simplest ideas, expressed in the simplest way, are the most powerful.

CHAPTER TEN

LISTEN UP

I have a workout buddy who gives me constant grief about my profession. He waits for those times when I am trying to lift a little more weight than usual. I have just put up six repetitions, sweating and grunting, and am desperately trying to summon the strength to put up two more. At this precise moment, and in violation of all the rules of gym etiquette, he casually asks me: "So, Mac, how does that make you feel?"

His theory is that all I do for a living is sit in my office and ask people, "How does that make you feel?"

Reflective Listening

I find this particularly aggravating because it is close to the truth. I spend a lot of time asking people how they feel. This is the easy part. The difficult part is knowing how to listen for the answer. There is a set of techniques to make us better at this part; these techniques are often referred to as reflective listening.

I will describe the techniques, but first a word or two about their purpose. Considered in isolation, the techniques by themselves can sound as dumb as my weightlifter friend makes them out to be.

Helping People Make Sense of Themselves

Most people want to be understood. And accepted. I don't want to be understood and rejected; especially when I am angry or frustrated or sad, I want understanding and acceptance. Then sometimes I can make sense of why I am feeling this way.

Much of the thinking behind the techniques of reflective listening stems from a psychologist named Carl Rogers. Rogers had an optimistic view of human nature. He thought we were basically good creatures who could work out healthy solutions to our problems if we were just helped to accept ourselves. Rogers thought we got ourselves tied into emotional knots when we expressed our feelings and then proceeded to get our feelings stomped on. As children, we may sometimes feel angry or jealous or selfish; if we can grow up in an environment where we can express these feelings and still feel accepted and secure, then we are more likely to be able to move past these negative impulses to more positive ones. This is one of those theoretical questions that is hard to prove up or down. Certainly it seems reasonable that Rogers had a fix on at least a piece of the truth about human beings. And one good thing that grew out of Rogers' thinking was a set of specific ideas about improving relationships and improving communication.

When I talk with someone in my work, I am interested in more than the surface content of what they say. I am usually especially interested in discovering the feelings behind the words. Mostly people don't begin conversations by talking about their feelings. They often begin by talking about behavior—they tell me some of the things they have been doing. Then perhaps they tell me some of the things they have been thinking. To do my job well, I need to know about their behavior, their thoughts, and their feelings. It is the last that is usually the most difficult to get at. Sometimes people are just cautious in talking about their feelings. Or they may be embarrassed, especially if they have been stomped on in the past for describing their feelings. Or, as if often the case, they have genuine difficulties describing their feelings because their feelings are such a confused jumble.

Open Up the Conversation

Reflective listening techniques are especially useful for talking about feelings because they open up a conversation rather than shut it down. Mr. Bird comes in to see me with steam coming out of his ears. I ask him why he is upset. He says he is upset because his supervisor didn't like his work. I respond by lecturing him that he should work more carefully so his supervisor will not criticize him. This is a conversation that has been thoroughly shut down because I said the wrong thing.

With reflective listening, the trick is to reflect back to the other person the essential message he is trying to communicate. One way to think about this is to imagine an image that is being bounced back and forth in a hall of mirrors. In this case, though, the original image (or message) is distorted, and each time it bounces off one of our magic mirrors it gets a little clearer. We have to keep the image bouncing back and forth long enough that we can begin to see it clearly.

If done poorly, reflective listening just makes us sound like parrots. Done well, it improves communication.

Mr. Bird: "I hate my boss!"

Good Response: "You seem pretty mad."

Bad response: "You shouldn't let yourself get so upset."

Mr. Bird: "She is so dumb."

Good Response: "She must have done something you didn't like."

Bad response: "She can't be that bad."

Mr. Bird: "She made me do all my work over. She's an idiot."

Good response: "I hate it when I have to do stuff over."

Bad response: "You must have done the work wrong."

Mr. Bird: "I'm not going to get paid much because of her."

Good Answer: "Bummer, dude."

Bad Answer: "Well, you'll just have to work harder tomorrow."

Mr. Bird: "Yeah. And I was mad. Everybody else finished before I did."

Good Answer: "So you were mad because you had to do the work over. And bummed out because you're not going to make as much money. How did you feel about everybody else finishing before you?"

Bad Answer: "You can't always finish first."

Mr. Bird: "I felt stupid. Like I'm always the dumb one."

Good Answer: "I'm sorry. That's pretty discouraging."

Bad Answer: "Well, I think you're smart."

Mr. Bird: "Back in school, I was always the last one to finish. I hated that. That's why I do my work so fast. So I don't have to feel like I'm the dumb one."

At this point in our conversation, we are getting closer to some of the feelings behind Mr. Bird's anger. We are getting closer to the heart of the matter. There are many places in this conversation where we could have shut the conversation down, but reflective listening provides a way of opening up the conversation and allowing Mr. Bird to open up about himself.

It should be noted that a good answer is, in reality, anything that moves the conversation along. It might be saying nothing but showing through posture, through facial expression, an interest in what the person is going to say next. It might be saying yeah or huh or really? as long as these are responses that move the conversation down the road.

How to Get People Not to Talk to You

The bad answers given in the example above illustrate some common mistakes although there is no way to list all the possible mistakes that can be made when one human being tries to communicate with another.

One kind of bad answer is to be judgmental—I think there is probably a minor joke in here somewhere, but I will let it pass. We can save the judgments for later because in therapeutic conversations we are trying to help the person come to his own judgments.

We can fall into another trap when we confuse being critical with being helpful. Sometimes we rush to tell the individual how to behave because of our hope that this will cure the problem and keep him from ever feeling miserable again. Another kind of bad answer is to simply try to cheer the individual up. Sometimes we do this because Mr. Bird's misery makes us miserable, and we are trying to force him to be cheerful so we can be cheerful.

With the good answers, we are trying to communicate acceptance. We are trying to move things along so Mr. Bird can make his feelings clear to himself. We are trying to let Mr. Bird know that his feelings are important, respected. We are trying not to tell him how he is supposed to feel. We are trying, as much as possible, to provide a setting where he can think through his feelings and arrive at his own conclusions about them. We are providing a setting where he can talk his way through his own problem and arrive at his own solution.

Empathy and Detachment

One important skill in all this is to be able to separate our own feelings from those of the other person. As noted above, sometimes we fall into the trap of blindly trying to cheer someone up because their misery makes us uncomfortable. Being a good listener requires both empathy and detachment. I have to care while at the same time remembering that we are talking about Mr. Bird's problems and not mine.

In this game of problem ownership, it helps to use the word, I, in conversations. This trick separates my feelings from the other person's. It says to the other person: this is what I feel, but you do not have to feel the same thing.

Without sufficient detachment, it is easy to get tricked into volunteering to solve Mr. Bird's problems for him. If I find myself volunteering to talk to Mr. Bird's supervisor for him, so that we can clear up misunderstandings, I have lost my detachment and robbed Mr. Bird of the opportunity to solve his own problems. Another tempting mistake is to rush in and adopt Mr. Bird's initial view of the situation. Mr. Bird's first description of the problem suggests that the problem is very simple: he has a mean and unreasonable supervisor. If this is as far as Mr. Bird and I get in understanding the problem, then we haven't traveled far. The supervisor is not there volunteering to do things Mr. Bird's way, and Mr. Bird has found nothing to change since we have agreed that the problem belongs to the supervisor and not to Mr. Bird.

Problem Ownership and Solution Ownership

After Mr. Bird has talked about his feelings, he has to figure out what he is going to do. I have been a psychologist for lots of years, and I have spent thousands of hours listening to people describe some of the most confusing problems imaginable. I will be eternally grateful for the fact that it has never been my job to figure out what they should do about their problems. If this had been part of my job description I would have dropped from exhaustion years ago. I am pretty much overtaxed by trying to figure out what to do about my own problems, and the solutions I have managed to come up with have been about half-bad about half the time. I am certainly not smart enough to spend most of my working hours figuring out solutions to the problems of others.

Sometimes people come to me with the idea that I am going to tell them the solutions to their problems. Getting over this idea then becomes the first problem they have to solve.

The way to help people solve this and other problems is to ask them questions. This was a favorite trick of Socrates, and it is still reliable after all these years. A problem with Socrates is that he liked to lead people through a series of questions and at intervals along this garden path the Great Master would stoop to point out why a particular answer was just blindingly dumb.

Our goal is not to get people to thinking they are blindingly dumb; my personal observation is that life just naturally provides plenty of opportunities for me to have this experience of blinding dumbness without seeking out some Socratic advisor to pour salt on my wounds. I recall, for example, the recent occasion at the Wendy's drive-up window. I placed my order at the microphone and then drove dutifully up to the window. The teenager at the window told me how much money I owed, and I gave her a ten-dollar bill. She gave me back my change and said thank you. Having good manners, I said thank you, also. She closed the window. I put the change in my pocket and drove away. A block down the street I reached for my burger, realized it was not there, and realized further that my life had just been illuminated by another blinding flash of dumbness. What can I tell you? I had a lot on my mind. The teenager gave me something, my change, and said thank you and closed the window. It just naturally seemed to me that we were done, so I drove cheerfully away. When I noticed the small mistake I had made, I did a U-turn and drove back to Wendy's. The teenager at the window gave me my burger and drink and a look of pity that was touching in its own way, since she probably reserved this look most of the time for her own pathetic parents. As I collected my burger, I noticed the line of other teenagers at the inside counter casually stepping back so that they could peer out the window into the very face of Dumb. Some of them were trying to stifle their giggles, but I did not sense they were trying very hard. My guess is that there was not a skilled reflective listener in the bunch.

There is a gift to listening; there is a gift to opening up conversations; there is a gift to asking questions of people that will help them come up with their own solutions to their own problems. Fortunately it is a gift that can be cultivated. I have worked over the years on keeping my questions very calm and low key. I hope this calmness communicates that I have confidence the other person can come up with some good ideas. It also reinforces the idea that we are not talking about my problem here. If I get upset or tense or urgent, this can suggest that the problem is mine and perhaps I should be responsible for the solution. This calmness also suggests that we are simply interested in objectively sorting through the alternatives.

I ask questions of the following sort:

So, Mr. Bird, have you found anything that helps you get along better with your supervisor?

Do you think you are going to be able to make this job work out?

Do you think your supervisor is going to change and do things your way?

Do you want to keep this job or would it be better to look for another one?

Is there anything good about this job—perhaps the paycheck?

If you just quit, what would happen?

If you told off your supervisor, what would happen?

Have you ever tried telling your supervisor how you feel when she criticizes your work?

What problems does your supervisor face in doing her job?

Are there things you like about your supervisor?

Are there times when she is really nice and you have fun together?

And yes. For all of the above questions, I often add some version of "how does that make you feel." These are the kinds of questions that both open up a conversation and open up the person's thinking to a consideration of different ways of thinking about a problem. The odds are that with a consistent use of this kind of approach Mr. Bird will come up with some answers that make sense to him and that work for him. This approach does not always work. Winston Churchill described democracy as the worst form of government yet invented except for all the other alternatives. Asking people how they feel, listening reflectively to their answers, opening up conversations, defining problem ownership, calmly brainstorming possible problem solutions—none of these useful techniques guarantees a solution to either the problems of the individual or the problems of Western Civilization, but they provide some benefit, mostly do not hurt anybody, and, like democracy, are better than the alternatives.

And there is also this: Lynne, a most wonderful friend of mine, a psychologist, smarter than I am and therefore often aggravating to me, read this chapter and allowed as how it was mostly okay. She did made me take out a particularly dumb joke but let me leave the rest of the jokes in. She is doubtless smarter, but I can be a sly fox. Knowing my love for dumb jokes, she cannot bring herself to make me take them all out. I have therefore taken to writing chapters that include several poor jokes and one that is absolutely miserable. Even before she reads a chapter, I know which joke has got to go. After doing her duty with the jokes, my friend made one other substantive criticism of this chapter. Lynne said: "Mac, you spend too much time talking about technique. Technique is overrated. It doesn't matter much what words you use or what questions you ask. What matters is whether you care about the other person and are truly interested in what is on their mind. If you are, they will talk to you." I think Lynne is almost certainly oversimplifying the matter. It does occur to me, however, that I have spent, over many years, hour on top of happy hour, doing nothing but talking to Lynne. As I think back on these conversations, I cannot for the life of me remember what techniques she used to keep me talking so. I can remember many of the things we talked about, I can remember always looking forward to our next conversation, and I can remember the fact that her interest in me always steadied me in mid-stumble. I stubbornly continue to think she is oversimplifying, however, and I fully intend to tell her so on the very next occasion when I am lucky enough to be listened to by Lynne.

So Just Remember This

> ➢ **Reflective listening techniques are designed to help people express the feelings that underlie the surface content of what they say.**
> ➢ **The goal is to open up communication rather than shut it down.**
> ➢ **It is more than just parroting what someone has said; reflective communications can be anything that expresses acceptance and empathy.**

- ➤ We are trying to provide a setting in which the individual can think through his feelings and arrive at his own conclusion.

- ➤ It is good to separate out our own feelings from the other persons, so that we can communicate both empathy and detachment.

- ➤ The best solutions are those that the individual arrives at himself.

- ➤ We have to keep clearly in mind who owns the problem. These techniques are a way to communicate our interest in and concern for the other person.

CHAPTER ELEVEN

INCHING UP THE HILL

Individuals with MR/ID learn slowly. They change slowly. I suppose this is not the biggest of news, but it is a notion worthy of more consideration than it usually gets.

It is especially a notion that needs to be impressed upon those who are new to the field. And those who are impatient.

I have known for many years a woman with developmental disabilities named Gretchen. From time to time, I picture her now in the commons area of my building. She is usually playing cards or a board game with some friends. Sometimes I sit and we pass the time of day. Gretchen usually holds my hand when I sit beside her. She tells me about her job and her apartment and her bowling. She laughs and rolls her eyes when I inquire about her bowling scores. I often linger in talking with Gretchen because she holds my hand and smiles at me. I have a weakness for a good smile, and Gretchen has a great smile. It is weathered, wise, and radiates contentment. It was not always thus.

Ten years ago Gretchen was a terror. Policeman, brave and tested, would hear her name over the dispatcher's radio and immediately call in sick. Gretchen liked to browse the shops on Main Street. She rarely bought much because she was poor, but she liked to browse, and once in a while she would have a question about an item for the salesclerk. Sometimes the clerk was not immediately available. If forced to wait too long, Gretchen would lift the clerk by her lapels (or whatever was handy) and hold her in the air so the clerk could more readily give Gretchen her full and undivided attention. Gretchen's usual practice, when she was through with the services of the clerk, was to wad up the clerk like a piece of paper and pitch her on the floor. This was about the time every policeman in town was listening to dispatch and desperately hoping that fifty Hell's Angels would start a priority riot so they would not have to deal with Gretchen.

Gretchen no longer beats people up. She no longer wads them up and pitches them in the trash. Slowly, imperceptibly, over a span of many years, she got better. She mellowed. Granted, I still see, once in a while, a gleam in her eye, and she seems to be scanning the room for a cute little salesclerk. But those days are gone. She has learned.

I am supposed to be the expert in these matters, and I don't have a clue what brought this change about. I can remember that we tried about everything under the sun. We tried a reinforcement program. We tried an anti-depressant to cure her "masked depression." We tried counseling. We got her into bowling. We got her a different roommate. The judge finally lost patience and tried a couple of nights in jail. As I look back, I am reasonably convinced that none of these things fixed Gretchen.

If anything fixed Gretchen, it was Gretchen. And time. Individuals with MR/ID often change slowly. Those of us who work with them have to reset our clocks, revise our expectations. Sometimes, when we are up to our armpits in crises with an individual, it is helpful to step back and remember the Grand Canyon was once a ditch.

At our agency, we had a weekly meeting called Triage. Various supervisors and agency staff sorted through incident reports and tried to figure out what to do about the individuals who were tying us in knots that week. Gretchen's name gradually dropped off the list. Having been at the agency for many years, the most interesting thing to me about Triage were all the names that stopped coming up. I could roam the halls after Triage and see person after person whose name didn't come up that week. People who used to tie the agency into one giant, collective knot, people who snarled up all our good intentions and snarled at us in the process—they were out there in the halls and in their apartments and on their jobs, but they were no longer the problem of the week. They got better.

There was a psychiatrist from a long time ago named Adler who used to talk a lot about inferiority complexes and striving for superiority. When Adler talked about striving for superiority, he was not talking about social climbing or striving to be better than someone else. He thought we all have a pretty basic impulse to get better, to improve those things about ourselves that we see as inferior. I see lots of people with developmental disabilities who are working on getting better. Many are not racing up the mountain; many are intensely frustrated and angry at the slowness of their own pace; but they are inching up their own hills.

It sometimes takes the perspective of years to see the progress. I try to tell myself to be patient and watch for it. Sometimes, watching Gretchen's radiant and weathered smile, I wonder whether I have climbed my own hills half as well. Apart from metaphorical hills, I also like to hike in the summer Colorado's fourteeners (the highest mountains in the state). As you gain altitude, you gradually get even with and can see over the surrounding mountains—first the "twelveteeners" yield, then the thirteeners, until finally from fourteen thousand feet you can see over all the surrounding peaks, and there are mountain tops that roll on into the blue distance. I have sat on top of some fourteeners and looked far, but sometimes Gretchen's smile suggests to me that she is seeing a view from a higher place than I have yet reached.

So Just Remember This

> ➤ Sometimes change comes slowly to those with (and without) developmental disabilities. Be patient. It is often worth the wait.

CHAPTER TWELVE

WEIRD ENVIRONMENTS

Dear reader, there may be those among you who have paused now and again, and simply let the rush of events swirl by. In those pauses, you may have taken a careful look around and realized that everything in life does not always make complete and perfect sense. For those of you who have <u>not</u> noticed this, please read no further because I have no wish to disturb your tranquility.

For those brave enough to read on, I would like to talk about some of those things having to do with the developmentally disabled population that make no sense but can't be readily fixed.

I knew a man named Morris who was 58 years old. Morris is an individual with moderate MR/ID. Morris was also a bad-tempered grump. Morris liked every day to be the same. He liked to come home from work and sit in his favorite easy chair and watch the news on TV. He liked to have one beer, not two, while he was watching the news. He did not like noise. He did not like it if someone had moved the ottoman on which he liked to rest his feet while he had his beer and watched the same news channel every night. When this routine was varied, Morris got grumpy. When there was too much noise in the house, Morris got grumpy. If someone tried to talk to Morris while he was doing his thing, he got...you get the point. Most of the time, when Morris got grumpy, he muttered to himself and got quietly steamed. When his grumpiness exceeded a certain level, Morris broke most of the furniture in the house. He then threw the broken pieces through the various windows of the house. Morris did not always remember to open the windows before he threw the furniture through them.

In short, Morris was a guy who liked his peace and quiet and would smash everything up pretty thoroughly if he didn't get it. You gotta love this guy.

For most of the time I have known Morris, he has lived in a group home. Because he smashed things up on occasion, Morris was considered to have "behaviors." I note in passing that this is one of those bits of social science jargon that we could live happily ever after without. Everyone has "behaviors" except the deceased, so to describe someone as having "behaviors" does not add a lot of useful information. Because Morris had "smash everything up" behavior, he mostly has lived in a group home for individuals with MR/ID who also have emotional and behavioral difficulties. Usually there were six other residents along with a couple of staff. This made for a pretty crowded house, and crowded houses are most generally not quiet and peaceful houses. Most of the other clients who have lived with him were younger folks who were generally in their twenties. People in their twenties generally do not cherish peace and quiet. People in the twenties mostly cherish loud music. And they jump up and they jump down and around and can't sit still and they argue with each other and cannot have an argument without yelling and screaming at the top of their lungs.

I am happy to report that Morris now lives in a small home with two other developmental disabled individuals who are about his age and share many of his habits and preferences, meaning they are very quiet. Morris no longer breaks furniture. He is happier than I have ever known him to be. Who wouldn't be happy, having finally escaped all those noisy kids? Everyone knew it was crazy to have Morris living with all those loud young kids for all those years. I knew it. Morris knew it. His team knew it. Most importantly, the residential director knew it, but there was not a thing she could figure out to do about it.

The residential director was an administrator, a bureaucrat, and she was so good at what she did that these terms, when applied to her, are compliments. She carried in her head an elaborate chess board of who lived where, who got along with whom, who was getting ready to move to a more independent setting, how many people were on the waiting list for services, what the odds were that she could squeeze funding for two more slots out of the check-writers. If she got approval for one new slot, this could set into motion a series of moves and possible combinations of residents that boggles the mind. At least it boggled mine.

It took her several years to figure out how to get Morris the residential placement that everyone knew he needed. This is not a criticism of her slowness. I would still be trying to sort it all out.

During that several years, Morris broke a lot of furniture. Both Morris and the people that worked with him were frustrated and miserable a great deal of the time.

All the time Morris was breaking furniture, he was living in a weird environment, a setting that made no sense for him and made no sense to all the people that worked with him. It happens. The world is not yet perfect and does not yet make complete and total sense. Everybody tried their hardest, and it still took years.

In this business, probably in most businesses, we get confronted fairly regularly with situations that do not make complete and perfect sense. Dropped into the raging middle of such situations, we often end up like the drunk looking for his lost keys under the lamppost; he knows this is not where he lost the keys, but it is the only place with enough light to see. I did my part under the lamppost with Morris. We worked on anger control. Never have I had a more motivated client. Morris hated it when he lost his temper. He practiced relaxation exercises. He took relaxation tapes home with him and listened to them faithfully every day. He took lots of long 'walks. He rehearsed mental tactics for avoiding angry outbursts. He practiced conflict resolution with the other clients in his crowded home. He sometimes wore headphones and listened to music to shut out the noise of his fellow residents. He paid on a proportional basis for some of the damage he caused although his income was not high so he never came close to being able to afford the real cost of the damages. Maybe all this reduced the frequency of incidents—I don't now. I know I did my part; others on Morris' team did their parts. Everybody loved Morris, and everybody, most especially including Morris, tried as hard as they could to find the keys under the lamppost. Finally, the residential director solved the problem by finding the right placement for Morris, and he has lived happily ever after.

I keep trying to find the moral to this story, and it keeps eluding me. Perhaps I am looking for the moral under the lamppost, and it is lying out there beyond the range of my rather dim lights. I do know that everyone on Morris' team knew the situation was crazy, and no one on his team quit trying just because the situation was crazy. This is another way of saying that everyone on Morris' team really was on his team, and this made up for a little craziness and a lot of broken furniture.

And there is another thing. The guys at the glass company were regular visitors. They came to refer to it as "making the Morris run." They were kids (to me and to Morris) in their twenties. They themselves were a pretty noisy lot; they were extremely impressed with the number of windows Morris could break in the course of a rampage. Furthermore, they didn't know from Godzilla about respectful language, so they referred to Morris as a "kid" or a "boy" as they installed the new window panes. When the bills came in, however, there was a charge for the replacement glass, but under the column for labor charges there was always a zero with the accompanying notation—"the Morris run."

So Just Remember This

➤ Sometimes individuals get stuck in settings that make absolutely no sense for them.

➤ Everybody knows it; everybody wants to change it; but, sometimes you just got to hang in there until the world makes more sense.

CHAPTER THIRTEEN

FAMILIES

I approach this chapter as one about to skate on thin ice. I want to talk about families, and I've got the good intentions, but I doubt my skating technique is sufficiently sure to keep me safe. I may mess this one up because I do not have a family member with developmental disabilities. So, as I lace up my skates, it is already clear that I can have only the most limited understanding of what I am about to talk about, even though, I freely grant, this has seldom stopped me before.

There is a thing I would make clear from the outset: I don't want to know what it is like to have a family member with developmental disabilities. The thought of having a child with MR/ID frightens me. Already I can hear the ice cracking beneath me. I have written about the joy of my associations with individuals with MR/ID, about their courage, humanity, lives full of challenge and grace. But I do not want anyone in my family to have a developmental disability. I want everyone in my family to be from Garrison Keillor's mythical world of Lake Wobegon, where all the children are above average. So, the fundamental contradiction: individuals with MR/ID are wonderful, but, please, not my family. I expect this damnable admission assigns me pretty permanently to the human race.

The Dreams of Parents for Their Children

I don't know how families do it, how parents manage their lives, their emotions, in the face of this fix. Certainly there is a sense in which all children are perfect children, but some of these perfect children may not learn to read or develop street safety skills or make much money or establish "regular" relationships in which one gets married and has children and enters into the weave of life in the way in which parents hope. Few of us precisely fulfill the dreams of our parents, but it seems more love, more strength, might be required of parents of children with MR/ID. I don't know. You would have to ask such a parent to find out.

So, ask one.

I did ask one, a wise and old friend. I hasten to add that the adjective, old, applies to the friendship and not to my friend's age. I am not all that certain of her chronological age, but her spirit is that of a springtime chick. Neither am I all that certain of how long we have been friends—I think perhaps it seems we have been friends forever simply because I do not think we will ever stop being friends. In any event, I asked her about a draft of what I had written. In the margin, she wrote, "not more love, only more sucking it up and less whining." My friend gets testy if I try to admire her love, her courage. But let her get testy, it's my book, and I can admire that which is admirable about her, and if she doesn't like it, she can write her own damn book.

Invite Parents (and Siblings) In

We know what to do with parents of young children who live at home. Mental health types are eager to become partners with this class of parents. We take a history from them. We ask them what the problem is. We ask them what we should do about the problem. We often depend upon the parents to implement the solution to the problem. All other things equal, we assume they know their child best.

Our interest in parents drops off dramatically when their child does not live at home. This is mostly true whether the offspring is a child or an adult. Sometimes we lose interest in the parents. Sometimes we become antagonists. Sometimes we start blaming the parents for whatever is wrong with their offspring. There can be good reasons for this. There are some lousy parents out there. There are also some lousy psychologists, social workers, developmental disabilities specialists, etc.

Beware of Finger-Pointing

Caution is mostly in order when we get into the finger-pointing business. This is illustrated by the sad story of Bruno Bettelheim's theory of autism. Bettelheim likely did not set out to make parents feel miserable, but that turned out to be his destination. He started out to develop a theory of autism. Children with autism often have impaired emotional relationships; they often seem coldly indifferent to others. They often also have impaired communication, MR/ID, and behavioral oddities like whirling objects in front of their eyes. They are generally children who are severely impaired in many aspects of their functioning.

Bettelheim thought (inaccurately) that these children were mostly born to very intelligent mothers. He developed the idea that these mothers were emotionally cold, emotionally distant, and really did not want their child. The children sensed their mother's rejection and became emotionally impaired. There was no great rush of mothers of children with autism lining up at Dr. Bettelheim's office, eager to tell him that they were emotionally cold and rejecting. Certainly most, if asked, would have told him exactly the opposite. This did not deter Bettelheim. Influenced by Freud, he concluded that these mothers were underconsciously rejecting their children and thereby causing their autism. This is the kind of brilliant scientific argument that can leave a parent feeling pretty cruddy. And, of course, it turns out that Bettelheim's ideas were neither brilliant nor scientific nor true. In Hamlet, Horatio thought of himself as a man more sinned against than sinning; we now know that parents of children with autism are more rejected than rejecting.

Learn From Parents

So, if we can't ask Dr. Bettelheim (he is deceased), who can we ask? We can ask a parent. And after asking, we can shut up for a while and listen and preferably take some notes. What is your son's favorite food? Does he sleep well? What makes him happy and sad and mad and nervous and scared? What helps him have more of the happy moments and fewer of the other kinds of moments? What are his weaknesses? In what areas do you want him to improve?

We can make parents full partners. Most parents are excellent advocates for the needs of their children. (Neither should it be overlooked that they can also be excellent advocates for the needs of service providers.) As advocates for their children, they sometimes complain about services and yell at us when we don't do a good job. We work in systems that are chronically under-funded, and we work with problems that are often too difficult for us to solve. Here is one important principle to keep in mind: <u>of course, we are going to screw up.</u> How could it possibly be otherwise? And the corollary to this is that parents are also going to screw up sometimes.

Over the years, with much trial and error, I have tried to learn a system for dealing with family advocates who are yelling at me because I screwed something up. I say—this technique gets complicated so pay attention—"I'm sorry. I messed up. What can I do to make things right?" I also try to recognize that I would likely be yelling the same things were we talking about my son or daughter or sibling.

Sometimes Sorrow Surfaces

Also, I try to keep in mind that some of the yelling may have to do with my screw-up and some may have to do with anger and grief that does not go away for the very simple reason that MR/ID does not go away.

In a novel called *The Hotel New Hampshire,* John Irving writes about a Labrador retriever named Sorrow. Sorrow was such a faithful companion to his owners that they had him stuffed when he died, so they could always keep him close. If all this sounds too sorrowful, I should quickly point out that Mr. Irving fills his book with characters (even including Sorrow) that are eccentric but full of life and energy and joy. What other kind of characters would think of stuffing Sorrow? After Sorrow is stuffed, he has the misfortune to be placed on a transatlantic flight that crashes. After the plane goes under the waves, one of the things that bobs to the surface is Sorrow. I have asked parents, and they tell me that life is often full of joy, but this does not change the fact that sorrow, like Sorrow the Labrador, may bob to the surface at unexpected moments.

I have mostly found it to be a good idea not to expect parents to behave or think or feel in any particular way. Some want to be actively involved in every aspect of their offspring's lives. Some want to withdraw from such close involvement. And sometimes it is the same parent who at one time wants active involvement and at another wants to withdraw. What sense is to be made of such contradictory behavior? I don't know, but I can pass on some part of the picture I have gotten from asking parents.

Parents Can't Switch Jobs

Parents cannot resign their position. There is a lot of staff turnover among those who work with the developmentally disabled. This is due to a whole bunch of things like high stress, low pay, etc. Staff get burned out and quit. Parents sometimes get burned out; they cannot so easily quit, but sometimes they need to withdraw for a while. So, let them. Parents may need a sabbatical on occasion in order to return more refreshed to this part of their lives.

Parents Have Seen the Whole Movie

Keep in mind that parents have been there pretty much forever. Mr. Barkley is 35 years old and has severe MR/ID and resides in a group home. Mr. Barkley's parents have been around for all of those 35 years. Mr. Jordan works an eight-hour shift at Mr. Barkley's group home and has done so for the past three months. Three months is a pretty small amount of time compared to 35 years. I just fetched out my calculator, and 35 years times 12 months equals 420 months, which, if my math skills are still intact, is quite a bit more than three months. As a matter of fact, Mr. Jordan's experience with Mr. Barkley represents .007 per cent of

Mr. Barkley's parents experience with Mr. Barkley. Parents might consider bringing a calculator with them to meetings just to sort of level out the playing field.

Mr. Jordan has a snapshot of Mr. Barkley, but Mr. Barkley's parents have sat through pretty much the whole movie. It takes a long time to get to know individuals with MR/ID, longer than for individuals with higher IQ's. It may seem it should be otherwise, but I am convinced it is not. So, as paid staff, as short-timers, if we are confused—and we are often likely to be confused—it is a good idea to ask questions of those who came in at the beginning of the movie.

In Families, People Depend on Each Other

The fact that parents cannot resign their positions creates what in my trade are referred to as dependency conflicts. One of my pet peeves has to do with the amount of crap having to do with dependency questions that is passed off in my profession as if it were divine wisdom, when it is actually more on the order of passing gas. We are social creatures. We are family creatures. We depend upon each other. This is not a disease.

Sometimes when mental health types talk about dependency, and most especially the dreaded "co-dependency," it begins to sound like a sickness. Sometimes, I suppose, it is. Spouses who suffer themselves to get beaten up over and over by alcoholic husbands have a problem. It's a matter of degree and context.

I think families are the particular place where we all try to work out for ourselves how much we are going to depend on others and how much we are going to allow them to depend upon us. This is a central problem in living but not a sickness. It is a problem in living that allows for more than one right answer. Some families "co-depend" on each other a great deal; others, not so much. Those who do not "co-depend" at all are not families in the sense I understand the term; I am only completely independent of those I have no relationship with.

The family is generally the place where all of us work out our positions on these dilemmas because families usually include at some point some helpless members. Teenagers are, of course, the most knowledgeable and independent members of the family; at the other extreme are the relatively helpless—infants, young children, middle-aged fathers. In one version of "co-dependency," the wife "enables" her husband to drink and be abusive. In another version, however, the parent "enables" the child to eat and grow and be safe and learn and love. Parents are perpetually baffled as they try to work out how much help to give, when to comfort, when to ignore, and when to kick a little butt (figuratively speaking, of course).

Problems of this sort do not belong exclusively to psychology or to science and do not have simple right or wrong answers. These questions also belong in the realm of values and religion and philosophy. When do we promote self-reliance in others and when do we cross the road to help like the good man from Samaria? When do we forgive seven times seventy times and when do we get fed up and drive the money-changers from the temple? To what extent are we our brother's keeper?

When we are parents, to what extent are we our children's keepers? And what is the answer to this question when our child is thirty-five and mentally retarded? These are dependency conflicts, relationship conflicts, and I do not know the answers to these questions; I simply know that families of individuals with MR/ID are perpetually plopped in the middle of these riddles.

And, similarly, the individual with MR/ID is bound up in the same struggle to find a balance between depending and "independing." This struggle is more acute when we consider the notion of mental age *vs.* chronological age. For a number of very good reasons, the idea of mental age is an over-simplification that should always be avoided except for this one time when I need it to make a point. Consider the adult with a mental age of five. In some ways, he keeps on being five over and over as the years go by. This can be like putting a stick in the bike's spokes; the ride is smooth and happy and then for no apparent reason the wheel seizes up, and the individual, and often his whole family with him, is tumped abruptly into the ditch.

Every family has their own way of getting out of the ditch. Lacking perfect answers, all I know to do is to try to have good manners. Lend a hand when it is asked for and not otherwise. Listen to this family's story. Be respectful of its grief and delight in its joy. I try to offer encouragement when I can, but it is not my place to approve. I try to offer as much understanding as I have in me without necessarily thinking it is all that much. I try to keep in mind, as simply a blind article of faith, that most people, most of the time, are as strong as they need to be.

So Just Remember This

➢ We should take pains to include the families of individuals with developmental disabilities in our efforts.

➢ We should be careful when we are tempted to blame parents for the problems of their child.

➢ Parents and siblings often make excellent partners in advocating for the needs of the individual.

➢ Parents may experience recurring grief and stress over the problems of their child and may sometimes need a vacation from their efforts.

➢ Parents usually have a much longer and deeper understanding of their child than we do, so pay attention.

➢ As with most families, parents are likely to have emotionally complicated relationships with their child. It is often a continuing struggle to work out the boundaries of relationships, especially as regards dependence and independence. We should be respectful of those relationships, about which we are likely to have only a limited understanding.

CHAPTER FOURTEEN

SWIM OR SINK

Working with individuals with developmental disabilities often involves high stress and non-high pay. There are abundant rewards, but often they don't show up on a W-2.

I don't know what to do about the low pay, but I have some tips about coping with the stress of the job.

You Can't Save Your Clients from All Pain

I know a wonderful married couple, both of whom are individuals with mild MR/ID. Bob and Sally are devoted to each other. I do not know any couple more attuned to the needs of the other and more considerate of those needs. It is a joy to watch them in a crowded room when they are separated, talking to others. Every so often one or the other of them looks up and finds the other with a glance, and they nod to each other. Both are far too shy to wave or shout or even smile, but the nod is always there. The nod is grave, deliberate, and to my eye, knowing them as I do, it has the grace of two swans bowing to each other across a still pool.

Bob and Sally do not resemble swans in all ways. Sally has a bad hip and a pronounced limp. Bob has tuberous sclerosis, a condition that results in large bumps on his face. One day I was trying to describe for them directions to a coffee shop they might like. I knew they often took walks around town. I explained a route that would take them down Fourth Street. They had been paying close attention, but at this point in the conversation their faces turned into masks, and both seemed to withdraw from the discussion. I asked what was wrong. Sally waved her hand dismissively and said that they did not walk down Fourth Street. I was baffled and asked why. I hastened to point out that it was within easy walking distance of where they lived. Patiently, kindly, Bob explained to me that there were always kids on Fourth Street. "They call us ugly. They call us retarded." I expressed my anger, and Bob gently put his hand on my shoulder to comfort me and reminded me that these were just kids. The picture of the two of them making their way down Fourth Street has stayed with me. Bob is a big, slow, lumbering man, and Sally's hip makes her always seem to be pivoting around some obstacle, and they would not be able to speed up and escape this meanness; they would be able only to swim slowly through it and perhaps exchange one of those grave nods when finally they were safely away.

Unfortunately, no one can save Bob and Sally from this pain. Fortunately, they are as strong as they need to be and are well able to help themselves. The helping professions often attract people who want to fix things for others. This is as it should be, else why would they call it helping? But there are limits on what can be fixed, and failing to recognize these limits can make the helper pretty miserable.

Leave Work at Work

When I first started out in this business, I took office worries home with me. I had trouble getting out of my mind the problems and occasional tragedies of the day. I am not the world's fastest learner, but I eventually came to realize that taking these worries home did not help my clients and did not help me or my family. This realization, in itself, was not enough to fix the problem. I eventually found that exercise was helpful. Every day after work I go work out for an hour in the gym. This breaks up the thought patterns about work. By the time I leave the gym I am more able to focus on other things besides work.

Worry Doesn't Help

This advice comes from one of the world's most accomplished worriers. Thinking about a problem sometimes helps, but worry and anxiety do not. If, like me, you have to worry, then set aside some time every day to spend worrying, and limit your worrying to those times. Sometimes it helps to make a list of your worries. Then stick it in a drawer and forget about it. Or, if you are an incredibly productive citizen, you can write down beside each worry what you can do about it. Even more useful, for most worries, you can write down that there is nothing you can do about it so you might as well stick it in the drawer and forget about it. None of us can get by without some worries, but it is a good trick if you can figure out ways to limit the time you spend with this troublesome companion.

Many of the things we waste time worrying about involve outcomes that are not the end of the world. So take the worry to its extreme conclusion. If I do this, and I discover that the worst possible outcome is that someone will think I'm stupid or someone won't like me, I begin to grab a clue that the thing I am worrying about may not be the end of the world. For these kinds of worries, it helps me to remind myself that the worst possible outcome is one that I can still cope with.

Learn to Count

Sometimes, at the end of the day, I sit back and take stock. In nice weather, I sit on my deck in that time between sunset and dark and watch the colors of the western sky shift from orange to purple to gray. During this calm and serene time, I think back over my day and beat the hell out of myself for anything I did wrong that day. This happens because I am poor at math. Most folks have this particular kind of mathematical inadequacy. We count the one screw-up of the day and forget the 20 or 40 or 100 things we did right that day. Ted Williams may have been the greatest hitter in baseball, but even at his best the pitcher still got him out about six times out of ten. Michael Jordan was happy when he only missed about half his shots. And then there I am, watching the afterglow of the sunset, counting only my mistakes, and then multiplying them by 50 so I can work my depression up to a fever pitch.

If we are going to keep score, and let's face it, we are, then we can at least count fairly. There is no law that says we cannot count our successes. There is also no law that says we cannot celebrate the small things we did right that day.

Think Small

In my youth, I had every intention of saving the world. I still think I may get around to it (saving the world, that is), but between skiing and fly fishing and the unreasonable demands of work and family responsibilities, I've been pressed for time. Maybe Friday afternoon I could get around to it, although there is a leaky faucet in the kitchen that really needs my attention. Grand ambitions are certainly grand, but sometimes the big picture can distract us from the small. The advantage to the small picture is that it can be made specific. With the small picture, I can know when I have had a successful day. Did I get up in time to catch the sun coming over the mountain while I was shaving? Was I able to forget everything else and really focus on what my clients were saying to me? Did I stay up on my paperwork so it does not pile up for the weekend? Did I manage to come up with at least one useful idea for somebody? More importantly, did I manage to find a way to present it in such a way that they could accept it and put it to good use? Did

I get in my quota of exercise? I love that endorphin high. Did I take an extra moment with someone to remind them how special I think they are? Was I able to laugh and make somebody else laugh so there was some fun in the day? In spite of my screw-ups, did I remember to give myself credit for all the things I did right that day? When the goals are this small and specific, I can know whether I had a good day or not. By thinking small enough, I can practically guarantee myself lots of happy days. I have a lurking suspicion that saving the world might not be such a sure thing.

If It Hurts When You Think That, Don't Think That

I love that old movie where Groucho Marx is the doctor, and the patient comes to him with an arm that hurts when he moves it a particular way. The patient demonstrates the movement that hurts, and says, "Doctor, it hurts when I do this." Groucho waves his cigar in the air and says, "Well, don't do that!" Often, in therapy, I relay this story to my clients and do my best Groucho imitation. Unfortunately for my clients, I love this Groucho routine, and I do it terribly. If Groucho ever saw me doing Groucho, he would say, "It hurts when you do that." Apart from my love of this little routine, the reason I am so often relaying it to my clients is that they so often are thinking things that hurt, and they need to learn how not to think that.

They think:

I am not lovable.

I have to be loved by everybody.

I am not competent.

I have to be perfectly competent.

If things do not work out just so, it will be a complete disaster, and the world will probably come to an abrupt end, and it will be all my fault.

I never do anything right.

I am not strong enough to handle this situation.

I am no good.

I don't deserve to be happy.

Everything is beyond my control.

Everything is in my control, and I have screwed everything up (or soon will).

The only reason I haven't screwed everything up is that I am so terribly inefficient.

These are all recipes for misery. I note that this is not the complete list; I am suspicious the complete list could go on forever. We are all prone to this kind of thinking on occasion. It is extreme thinking; it is all-or-nothing thinking. I have become convinced that we should never allow our children to say the word never. And we should always forbid them to use the word always. Our children say to us, "You never let me do ....," or, "You always let my brother do ...." (those of you with children can fill in the blanks without difficulty). This is childish (and manipulative) thinking. It is illogical because it leaps from one specific incident to a general statement. Mr. Spock would condemn it as illogical; Dr. Spock would understand it as childish.

We can't avoid getting older, so most of us do the next best thing and avoid getting wiser. We often continue to think in these extreme ways. "If I am not perfectly competent, then I must be perfectly incompetent. If I make one mistake (or even two or seven), then I must be a complete screw-up."

We would be better served if we got up every morning, looked in the mirror, and said, "Mirror, mirror, on the wall, who's..." No, no, that's not it. We should say, "I am partly lovable, and partly smart, and partly strong, and partly honest, etc., and anyone who says otherwise is partly full of ga-ga."

If You Insist On Self-Improvement, for God's Sake, Go Slow

Actually, you can go as fast as you want. But keep in mind that most change happens slowly. I do my Groucho routine and say to people that if it hurts when you think that, then don't think that. They eventually get around to asking me how it is exactly they are supposed to stop thinking "that," and this is the tough one.

Freud thought you could stop thinking "that" by having an insight. For Freud, you got insights in therapy. Freud liked insights about what happened to you in childhood. He was especially interested in traumatic occurrences in childhood although he defined trauma pretty loosely relative to the way we use this term today. Assume, for example, that when you were five your father always forbid you to say always and never allowed you to say never.

This might be pretty traumatic; it might be so traumatic that you might repress it—you would never remember what you always forgot. In therapy, Freud would help you dig up this buried bone, whereupon you would realize that the reason you keep thinking things that hurt to think about is because of this childhood event. This realization is an insight, and experiencing this insight should allow you to follow Groucho's advice and live happily ever after.

The only flaw in this plan is that it doesn't work. Usually insights alone are not enough to produce change. And most people do not need an analyst to help them figure out how they got to be the way they are. Mr. Smith comes into my office and complains of chronic feelings of inadequacy and depression. I inquire as to the origins of these problems. He replies that for 30 years his parents have been assuring him that he is a worthless piece of ga-ga, who will never amount to anything. Mr. Smith does not require my godlike powers of deduction to figure out why it hurts.

He wants to quit thinking this way, and this is the hard part. If an insight alone won't do the job, then it's going to take practice. He has had 30 years of practice thinking one way. Mr. Smith is like a record player, and every time you shake him a little the needle falls into this one groove on the record and gets stuck and goes round and round. He has to practice picking up the needle and putting it down in another spot on the record. And the key word here is practice. This is usually about the time I remind my clients of the Carnegie Hall joke. As everyone will recall, a tourist in New York City stops a native on the street and asks if he can tell him how to get to Carnegie Hall; the native replies, "Practice, practice, practice." I love that joke almost as much as the Groucho routine.

When you practice, you have to start small. And it helps to think small so that you are not trying to solve all the world's problems in one day. And it also helps to reward yourself for the small things you do right, and you can't do this very well if you don't learn to count accurately.

Don't Get Mad, Get Smart

I suggested in a previous section that anxiety was a problem in which we usually think too much, and all that thinking doesn't help. Anger is a problem that is helped by a little more thought, not less. Or, at least, some different thoughts. Anger is usually thought of as a response to frustration, and this is certainly a reasonable notion. But we know that aggression occurs in response to other things besides just frustration, and we also know that frustration does not always lead to aggression. Somewhere in the middle of frustration and aggression is some thinking, the stuff that goes on in our head in situations involving the possibility for anger.

A psychologist named George Kelly defined anger as an attempt to force validation of a construct. This is pretty powerful, big-word, shrink talk. Another way of saying almost the same thing is that we get mad when the world doesn't act right. We get mad when things don't go as we planned. We get mad when our ideas about how things should be don't work; when this happens, we try to bend others to fit our ideas rather than bending our ideas to fit others.

I once worked with a man named Terry on his problems with anger. I quickly diagnosed his problem; we developed good therapeutic rapport; we had weekly sessions in which we had spirited and intense discussions; and, at the end of our work together, Terry went back out into the world an unchanged man. Terry came in mad and left mad. At least I didn't make him worse. I certainly tried not to aggravate him unduly as he was a big, mad guy.

Terry was an individual with moderate MR/ID and a pretty bad seizure disorder. Terry approached interpersonal relationships with one great article of faith. If people around him were not acting right, they should be nudged towards the path of righteousness, and the best technique in the area of righteousness-nudging was to whomp them one. Or two. They should be made to act right. Terry encountered a number of individuals in his vocational program who did not always act right. I worked with Terry on letting staff handle these problems. When I suggested this approach to him, a light came to his face, he saw the promised land, and a weight was lifted from his shoulders. I sent him away, and I was content and well pleased with my great powers. When Terry came back the following week, he was angry and told me that he had had to whomp three more people. I asked him what had happened to the idea of letting staff handle these problems. He shrugged his shoulders, baffled, disappointed. He said: "I tried what you said, Mac. I left it up to staff." I said: "So, what happened?" He, a man betrayed, said: "Staff never whomped anybody. All they ever did was talk to them."

Terry was a man with an idea. Unfortunately, Terry was a man with only one idea—when someone's behavior is in error, you whomp them until they act right. Eventually Terry resigned as my client. He did this more in sorrow than in anger, persuaded at last that I was just one of those people who were never going to get it.

In dealing with situations that make us angry, it is good to be a person with more than one idea. Terry had a variety of good reasons for being a man with one idea; these included a harsh and punitive childhood and the likelihood of some brain damage that made it very difficult for him to develop any flexibility in his thinking. Working with individuals with MR/ID can often make us angry. It is always good to ask ourselves if we are adopting the Terry solution. Are we trying to force validation of our pet idea? If we are constantly in conflict with a client, that should be our first clue that we need to try something different. Most often, we get in these binds because we are trying to teach someone a lesson.

Teaching people lessons is a dangerous idea. Helping people learn works better. The first involves forcing validation of a construct; it involves anger; it involves forcing our ideas onto others. The second approach requires flexible thinking. It asks us to drop our reflexive responses to "angry" situations. If a client screams at you, he probably expects you to scream back—consider the possibility of simply listening quietly until he has screamed himself hoarse.

We all need to watch ourselves for signs that a situation or relationship is making us angry. Do we think about it constantly and get angrier and angrier? Can we can think of only one way to respond and that is with anger? I don't know whether rigid thinking induces anger or anger induces rigid thinking, but I know that the two of them are best pals, and neither of them is a friend to solving problems.

One of the hardest things to deal with in this business is angry clients. Don't they realize we are trying to help? Don't they appreciate it? This isn't a very happy situation, but it can get tragic if staff let clients teach them how to be angry. Step back. Use flexible thinking. Brainstorm the problem with others who are not directly involved.

Remember Your Biology

You will be relieved to know this section is not about going to the bathroom; I assume you can remember to relieve yourself without benefit of assistance from me. You will be disappointed to know this section is not about sex; again, you're on your own. Having ruled out two pretty interesting topics, I propose to talk about the biology of stress.

I like a certain amount of job stress. If the job is too easy, too routine, boredom and my natural laziness take over, and I fall asleep at my desk. This does not inspire confidence when people wander in unexpectedly. Mostly when I complain of job stress, therefore, I do not really want there to be no stress at all. Mark Twain had a funny bit on our popular stereotypes of heaven. Twain noted that a common vision of heaven was one in which we sit around all day playing the harp and singing. Twain further pointed out that he was not acquainted with many people, no matter how saintly, who could tolerate this activity for more than, say, 30 minutes. But in our vision of heaven, we propose to play the harp and sing not for 30 minutes, not for an hour, but for all eternity. It is not clear what would happen in these circumstances if you fell asleep from boredom. Perhaps you might be transferred to a livelier place.

Rarely, however, do I have problems with inadequate stress in my job. Sometimes the challenge of the job is just perfect. Sometimes it's a bit more than I prefer. When there is too much stress, my biological functions begin to assert themselves.

No, not those biological functions. The biological functions I am talking about are things like heart rate and respiration rate and blood pressure and sweating and muscle tension. Most folks know that under stress these biological mechanisms increase their activity. We are most familiar with the extreme versions of this stress response. If we narrowly avoid a dangerous car accident, it is easy to see these biological mechanisms at work because they are so extreme—our hearts are racing, we cannot catch our breath, etc. The smaller, more subtle versions of this response, however, are usually the ones that get to us on the job.

Anxiety often accumulates over the course of the work day in small increments. Each increase is so small as to go practically unnoticed, but by the end of the day we realize that we are "stressed out." Knowing that anxiety builds up in frequent, small doses provides one key to combating it. I try to practice anxiety-reduction in frequent, small doses throughout the day. I leave a few minutes of intermission between appointments and meetings so I can take three minutes to put my feet up and close my eyes and slow myself down. Some people use breaks like this to meditate or practice deep breathing or muscle relaxation. These are all useful techniques. I prefer visualizing ski runs or trout streams, but to each his own poison. Again, whatever the technique, the key is to work frequently throughout the day on relaxation.

Nobody Is Holding a Gun to My Head

I chose this job. I can choose another. All things considered, and unfortunate as it may be, the main person responsible for my decisions is me. Personal responsibility is a real pain, and there are lots of time when I would prefer to avoid it. However, it helps me once in a while to remind myself that I am here because this is the path I chose, and while it is certainly pleasant to blame others for my difficulties, it does not accomplish a great deal. If I don't like my job, I am free to go looking for another although swapping jobs, like swapping spouses, is not necessarily a guarantee of eternal happiness—new jobs, new spouses, etc. sometimes come factory-equipped with their own headaches.

It is refreshing, in the same way that cold showers provide refreshment, to remember that we are in the middle of a particular swamp because we chose it. We can certainly choose something else. We can also choose to reinvent the job we have. We can learn some new skill to bring to bear on old problems. We can figure out some better way to sort the black buttons from the white buttons. We can use flexible thinking to improve some relationship that is driving us crazy. The possibilities are endless so the odds are pretty remote that we will run out of choices. None of us is boss of the world, but if we keep a clear fix on our choices, we can be boss of our own corner of our own street, and ain't power grand?

So Just Remember This

> ➢ **You can't save everybody from everything.**
> ➢ **Leave work at work.**
> ➢ **Worry is mostly not your friend.**
> ➢ **If you must grade yourself, at least grade fairly.**
> ➢ **Keep your goals small and specific.**
> ➢ **Beware of extreme thinking patterns.**
> ➢ **Allow yourself to improve slowly.**
> ➢ **Rigid thinking is anger's best friend—think flexibly.**
> ➢ **Find a way to manage the biology of stress.**
> ➢ **Take responsibility for your choices.**

End

Made in the USA
San Bernardino, CA
06 May 2015